# Introduction to Discrete Mathematics

# Introduction to Discrete Mathematics

KOO-GUAN CHOO & DONALD. TAYLOR

An imprint of
Pearson Education Australia

Pearson Education Australia Pty Limited
Unit 4, Level 2, 14 Aquatic Drive
Frenchs Forest NSW 2086

Offices in Sydney, Brisbane and Perth, and associated companies throughout the world.

Copyright © 1994 Addison Wesley Longman Australia Pty Ltd
First published 1994
Reprinted 1994, 1995, 1996, 1997, 1998
9 10 11 04 03 02 01

All rights reserved. Except under the conditions described in the Copyright Act 1968 of Australia and subsequent amendments, no part of this publication may be reproduced, stored in a retrieval system or transmitted in any form or by any means, electronic, mechanical, photocopying, recording or otherwise, without prior permission of the copyright owner.

While every effort has been made to trace and acknowledge copyright, in some cases copyright proved untraceable. Should any infringement have occurred, the publishers tender their apologies and invite copyright owners to contact them.

Designed by Don Taylor
Set in Computer Modern Roman 10/12 using TeX
Printed in Malaysia, CLP

National Library of Australia
Cataloguing-in-Publication data

Choo, Koo-guan, 1942–
  Introduction to discrete mathematics.

  Includes index.
  ISBN 0 582 80055 2.

  1. Mathematics. 2. Computer science – Mathematics. 3. Mathematics – Problems, exercises, etc. I. Taylor, Donald E. II Title.

  511.3

The publisher's policy is to use **paper manufactured from sustainable forests**

# TABLE OF CONTENTS

Introduction ... *viii*

## 1 Catalan Numbers — 1
Three more problems .................. 3
Problem Set 1 ..................... 5

## 2 Sets — 7
The algebra of set theory .............. 11
Venn diagrams ..................... 13
De Morgan's laws ................... 14
Summary ........................ 14
Problem Set 2 ..................... 15

## 3 Functions — 16
Brackets and railway wagons ............ 21
Summary ........................ 23
Problem Set 3 ..................... 23

## 4 Counting Principles — 25
Theory of the Multiplication Principle ...... 29
Summary ........................ 30
Problem Set 4 ..................... 30

## 5 Ordered Selections — 32
A combinatorial proof ................ 35
Summary ........................ 35
Problem Set 5 ..................... 35

## 6 Unordered Selections — 37
Selections without repetition ............ 37
Binomial identities .................. 39
Selections with repetition .............. 41
Summary ........................ 44
Problem Set 6 ..................... 45

## 7 The Inclusion-Exclusion Principle — 46
The general inclusion-exclusion principle . . . . . . . 49
Summary . . . . . . . . . . . . . . . . . . . . . . . . 51
Problem Set 7 . . . . . . . . . . . . . . . . . . . . . 51

## 8 Multinomial Coefficients — 53
Summary . . . . . . . . . . . . . . . . . . . . . . . . 56
Problem Set 8 . . . . . . . . . . . . . . . . . . . . . 56

## 9 Boolean Expressions — 58
Laws of Boolean algebra . . . . . . . . . . . . . . . 63
Summary . . . . . . . . . . . . . . . . . . . . . . . . 65
Problem Set 9 . . . . . . . . . . . . . . . . . . . . . 65

## 10 Karnaugh Maps — 67
Summary . . . . . . . . . . . . . . . . . . . . . . . . 72
Problem Set 10 . . . . . . . . . . . . . . . . . . . . 72

## 11 Logic — 74
Propositions . . . . . . . . . . . . . . . . . . . . . . 74
Quantifiers . . . . . . . . . . . . . . . . . . . . . . . 78
Logical puzzles . . . . . . . . . . . . . . . . . . . . 83
Summary . . . . . . . . . . . . . . . . . . . . . . . . 83
Problem Set 11 . . . . . . . . . . . . . . . . . . . . 83

## 12 Digital Logic — 85
Summary . . . . . . . . . . . . . . . . . . . . . . . . 88
Problem Set 12 . . . . . . . . . . . . . . . . . . . . 88

## 13 Mathematical Induction — 89
Summary . . . . . . . . . . . . . . . . . . . . . . . . 93
Problem Set 13 . . . . . . . . . . . . . . . . . . . . 93

## 14 Generating Functions — 94
Recurrence relations . . . . . . . . . . . . . . . . . . 97
Summary . . . . . . . . . . . . . . . . . . . . . . . . 99
Problem Set 14 . . . . . . . . . . . . . . . . . . . . 100

## 15 Linear Recurrence Relations — 101
Homogeneous linear recurrence relations . . . . . . . . 101
Summary . . . . . . . . . . . . . . . . . . . . . . . . 107
Problem Set 15 . . . . . . . . . . . . . . . . . . . . 107

| | | |
|---|---|---|
| **16** | **Formal Languages** | **108** |
| | Regular expressions; regular languages . . . . . . . . . | 110 |
| | Connections with counting . . . . . . . . . . . . . . | 112 |
| | Summary . . . . . . . . . . . . . . . . . . . . . . . | 113 |
| | Problem Set 16 . . . . . . . . . . . . . . . . . . . . | 113 |
| **17** | **Finite State Machines** | **114** |
| | Deterministic finite automata . . . . . . . . . . . . . | 114 |
| | Kleene's theorem . . . . . . . . . . . . . . . . . . . | 118 |
| | Non-deterministic finite automata . . . . . . . . . . | 119 |
| | Non-regular languages . . . . . . . . . . . . . . . . | 120 |
| | Summary . . . . . . . . . . . . . . . . . . . . . . . | 121 |
| | Problem Set 17 . . . . . . . . . . . . . . . . . . . . | 122 |
| **18** | **Grammars** | **124** |
| | Summary . . . . . . . . . . . . . . . . . . . . . . . | 128 |
| | Problem Set 18 . . . . . . . . . . . . . . . . . . . . | 128 |
| **19** | **Graphs, Trees and Catalan Numbers** | **130** |
| | Summary . . . . . . . . . . . . . . . . . . . . . . . | 135 |
| | Problem Set 19 . . . . . . . . . . . . . . . . . . . . | 136 |
| **20** | **Hints and Answers** | **137** |
| **Index** | | **145** |

# Introduction

This book has its origins in the lecture notes prepared in 1989 for a new Discrete Mathematics section of the First Year Mathematics course at the University of Sydney. In 1989 this part of the course comprised 19 lectures and in the original notes each chapter covered the material required for a single lecture. In rewriting the material for book form we have added a few new topics and many more examples but at the same time we have endeavoured to retain the original flavour of one chapter per lecture.

Discrete mathematics covers such a wide range of topics that it is difficult to give a simple definition of the subject. Whereas calculus deals with continuous or even smooth objects, discrete mathematics deals with things that come in "chunks" that can be counted. We will be a lot more precise about just what sort of "chunks" we are dealing with in the later chapters.

If your mathematical background is only high school calculus you could well believe that mathematics is only about numbers, functions and formulas for solving problems. If this is the case, the topics in this book may be quite a surprise because for mathematicians, computer scientists and engineers, Discrete Mathematics includes logic, set theory, enumeration, networks, automata, formal languages and many other discrete structures. That is what this book is about.

On the other hand, in 19 lectures we can only present an introduction to the subject and we must leave other important topics such as graph theory, error-correcting codes, discrete probability theory and applications to theoretical computer science to a second or third course.

The topics covered are set theory, logic, Boolean algebra, counting, generating functions, recurrence relations, finite automata and formal languages with a lot of emphasis on counting.

The set theory and logic is basic material which will be useful in many courses besides Discrete Mathematics. Counting problems which look quite hard when stated in ordinary English can often be solved easily when translated into the language of set theory. We give many examples that reduce to counting the number of functions of various types between sets, or counting the number of subsets of a set.

Boolean algebras provide a connection between set theory and logic and we give a brief treatment of this in the book. We also explore various ways of representing logical expressions using physical devices such as switching circuits and logic gates (used in computer chips).

A powerful method for solving certain counting problems is the theory of generating functions and the related theory of recurrence relations. This is covered in Chapters 14 and 15.

The last part of the course is an introduction to finite automata and formal languages. This relies on some of the set theory and logic covered earlier in the course. We also take the opportunity to look back at examples related to our favourite counting problem — the Catalan numbers.

**Some advice about the exercises**
Many exercises are stated in ordinary English and must be translated into symbols before a solution can be found. This process is inherently ambiguous. Often there is no unique "correct answer" to the problem: it all depends on how you interpret the words. What is important is to be clear about what *you* think the words mean, find a solution, and be prepared to defend it with a reasoned argument.

If you are unsure about the meaning of an exercise, discuss it with others. You may find that different people have quite different interpretations. A critical attitude towards this material will improve your ability to think clearly about mathematics in general.

Wherever possible draw diagrams. If the problem depends on a parameter $n$, work out the solution for some small values of $n$ before tackling the general case.

The more challenging exercises are marked with an asterisk. Chapter 20 contains hints and answers for most exercises.

# Acknowledgements

In developing the lecture notes which eventually became this book we were greatly assisted by our colleagues Geoff Ball, Sam Conlon and Philip Kirkpatrick who used the notes in their lectures. We also thank Diana Combe, Jenny Henderson, and Bill Unger for their valuable suggestions and their help with proof reading early drafts. In addition we express our appreciation of the help and encouragement of Ron Harper of Longman Cheshire in bringing this material to book form.

And finally, we are most grateful to Yit-Sin and Jill for their patience and understanding during the entire project.

K.G.C. and D.E.T.   June 1993

# 1

## Catalan Numbers

WE begin with the following problem, which we return to many times (and in many disguises) throughout the book. Firstly we state it informally.

*The Bracket Problem*

**Problem 1.1** *How many sequences of brackets are there in which each left bracket has a matching right bracket?*

We call such a sequence of brackets a *balanced string* of brackets.

*A 'string' is just a list of symbols written one after the other.*

A useful strategy to adopt when tackling problems like this is to look at some small special cases and then generalize whatever insights we gain from them. So we start by looking at the shortest possible balanced strings.

EXAMPLE 1.2 There is only 1 balanced string of 1 left and 1 right bracket: (). There are 2 balanced strings of 2 left and 2 right brackets: (()) and ()(). And here are the balanced strings of 3 left and 3 right brackets — there are 5 of them:

*The symbol ❑ signals the end of a proof or an example.*

$$((())), \quad (())(), \quad ()(()), \quad (()()), \quad ()()(). \quad ❑$$

Now that you've seen a few examples, how could you solve the following very special case of Problem 1.1?

**Problem 1.3** *How many balanced strings of 4 left and 4 right brackets are there? What are they?*

*Hint: there are 14 of them.*

You will probably find that it helps to be systematic about this. After all, you don't want to miss any strings and you must be sure you haven't listed any twice.

One approach is to list the strings in "alphabetical order". We don't usually think of "(" and ")" as being part of the alphabet, so what does alphabetical order actually mean? Well, we just extend the alphabet to include "(" and ")" and suppose that "(" comes before ")". (Alternatively, you could replace "(" by the letter "L" and ")" by the letter "R" and regard the strings as words in "L" and "R".)

Now try writing the balanced strings in alphabetical order. As a warm-up, begin with the 5 strings listed in Example 1.2.

The idea of systematically listing the strings of brackets in some sort of order is a very useful one. The effort of making the list should encourage you to think quite hard about the precise structure of the balanced strings themselves.

Using 1, 2, 3 or 4 pairs of brackets there are 1, 2, 5 or 14 balanced strings that can be made from them. The sequence continues

> *These numbers are known as the Catalan numbers.*

1, 2, 5, 14, 42, 132, 429, 1430, 4862, ...

and it is clear that making a list can become very tedious indeed. Rather than make a list we would like a *formula* for the number of balanced strings. In order to do this we must make Problem 1.1 a little more precise by introducing a symbol $n$ for the *number* of brackets. Then it becomes

**Problem 1.4** *How many balanced strings of $n$ left and $n$ right brackets are there?*

> *$c_n$ is the $n$-th Catalan number.*

Let $c_n$ be the number of balanced strings with $n$ pairs of brackets. (We define $c_0$ to be 1.) It turns out that there are many things that can be said about $c_n$. Here are some of them.

(i) $\quad c_{n+1} = \sum_{k=0}^{n} c_k c_{n-k}.$

(ii) $\quad c_n \sim 2^{2n}/n\sqrt{\pi n}.$

(iii) if $C(z) = \sum_{n=0}^{\infty} c_n z^n$, then $zC(z)^2 = C(z) - 1.$

(iv) $\quad \forall n, \quad c_n = \dfrac{1}{n+1}\dbinom{2n}{n}.$

In order to understand these statements we need to be able to read the notation. In order to *prove* them we need to explore some ideas from logic and set theory. That is what the rest of this book is about. For the moment we'll just define a few terms. Full explanations will have to wait until later chapters.

> **The sigma notation**

Equation (i) is an example of the $\sum$-notation (called the *sigma notation*). That is, we use

$$\sum_{i=m}^{n} a_i$$

as an abbreviation for $a_m + a_{m+1} + \cdots + a_n.$

**Recurrence relations**  Thus equation ($i$) (a *recurrence relation*) can be written in the form

($i$)'  $c_{n+1} = c_0 c_n + c_1 c_{n-1} + \cdots + c_{n-1} c_1 + c_n c_0.$

Recurrence relations will be dealt with in Chapter 15.

**Generating functions**  The power series $C(z)$ in ($iii$) is called the *generating function* of the numbers $c_0, c_1, c_2, \ldots$. This is explored further in Chapters 14 and 19.

**Binomial coefficients**  The symbol $\binom{2n}{n}$ which appears in ($iv$) is a *binomial coefficient*. In Chapter 6 we shall see that $\binom{m}{n}$ is the number of ways of choosing $n$ things from $m$ things.

**Asymptotic formulae**  The symbol $\sim$ in ($ii$) means that the ratio of $c_n$ to $2^{2n}/n\sqrt{\pi n}$ approaches 1 as $n$ becomes large. This is a consequence of Stirling's approximation: $n! \sim \sqrt{2\pi}\, n^{n+\frac{1}{2}} e^{-n}$. Nothing more will be said about this, but it is a fascinating topic and well worth exploring further after you've covered the material in this book.

**Logic**  The symbol $\forall$ which appears in ($iv$) is an abbreviation for the words "for all" or "for every". We shall study this in more detail in Chapter 11.

# Three more problems

The three problems which follow are all in some way related to the Catalan numbers. The precise connections will be given later in the book, but for now you might find it worthwhile to discover the connections for yourself.

**The Railway Wagon Problem**  **Problem 1.5** *There are $n$ railway wagons at $A$ on the track below. The wagons are moved from $A$ to $B$. It is assumed that the central track can accommodate all $n$ of them and that they travel only from left to right (i.e., they may not be moved from $C$ back to $A$). How many arrangements are possible at $B$?*

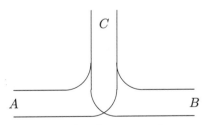

If the wagons are numbered $1, 2, \ldots, n$, the new arrangement at $B$ is called a *permutation* of $1, 2, \ldots, n$. We study permutations in Chapter 3.

EXAMPLE 1.6 Suppose that there are five wagons at $A$, labelled 1, 2, 3, 4, 5. Is it possible to produce the arrangement 5, 2, 3, 1, 4 at $B$? What about 5, 3, 2, 1, 4?

In fact it is impossible to produce the arrangement 5, 2, 3, 1, 4 at $B$. Here is an explanation.

In order to move 4 to $B$ we need to move 5 to the central track first. After that has been done, 4 is free to move to $B$ (via $C$). Next we must move 1 to $B$ but to be able to do this 2 and 3 must be moved to $C$ (below 5, which is still there). Now we have 1 and 4 at $B$ and 2, 3 and 5 at $C$. According to our rule that the wagons travel only from left to right, we cannot move 2 back to $A$ and so the only possibility is to move 2 to $B$ and then 3 and 5 to $B$. Thus the *only* arrangement at $B$ which ends in 1, 4 is 5, 3, 2, 1, 4. This explains why 5, 2, 3, 1, 4 is impossible.

In the course of the explanation just given, we have shown how to produce the arrangement 5, 3, 2, 1, 4 at $B$. ❑

*This is closely related to Problem 1.1.* Try counting the arrangements when there are 1, 2, 3, or 4 railway wagons. Do these numbers look familiar? What do they suggest?

The second problem is another type of bracketing problem, but this time we have symbols inside the brackets.

*Another Bracket Problem*

**Problem 1.7** Suppose you have numbers $x_0, x_1, x_2, \ldots, x_n$ and you want to multiply them together. In how many ways can you insert brackets into the string $x_0 x_1 x_2 \cdots x_n$ so that the order of multiplication is completely specified? Each pair of brackets should contain just two terms.

As was done for the bracketing problem, first look at some small cases. That is, try counting the arrangements when $n$ is 1, 2, 3 or 4. For example, when $n = 2$, there are two ways to bracket the expression: $(x_0(x_1 x_2))$ and $((x_0 x_1)x_2)$.

*For the answer, see Chapter 19.* This problem is closely related to Problem 1.1 but the connection is not quite as straightforward as you might think. Simply erasing the symbols $x_0, x_1, \ldots,$ and so on from their bracketed expression will produce a balanced string of brackets but this doesn't show that the number of bracketed expressions is the same as the number of balanced strings of brackets. Why not?

The third and final problem of this section is a first taste of formal languages and grammars.

*A String Production Problem*

**Problem 1.8** What are the strings produced by the following two rules and how many strings are there of length $2n$?

(1) $S \to \varepsilon$

(2) $S \to S(S)$

We have put a 1 or 2 beneath the symbol which is to be replaced. The number indicates which rule is to be used.

An explanation is in order. The symbol $\varepsilon$ stands for an empty string. That is, a string with nothing in it. To produce a string we begin with the *start symbol* $S$ and then use (1) or (2) to replace $S$ by the right hand side of the arrows, stopping when no $S$'s are in the string. For example:

$$S \to \underset{2}{S}(S)$$
$$\to S(\underset{2}{S})(S)$$
$$\to S(S(\underset{1}{S}))(S)$$
$$\to \underset{1}{S}(S())(S)$$
$$\to (\underset{1}{S}())(S)$$
$$\to (())(\underset{1}{S})$$
$$\to (())()$$

Perhaps you can see that this produces exactly the same strings that were considered in Problem 1.1. This sort of thing will be looked at again in Chapter 18 on Grammars.

There are many variants of these problems and others will be given in the problem set.

Eugène Catalan (1814–1894)

The numbers $c_n$ are called the *Catalan Numbers*. E. Catalan wrote an article about them in 1838 but they had been studied by J. von Segner and L. Euler 100 years earlier. They also occur in the work of the Chinese-Mongolian mathematician An Tu Ming.

An Tu Ming (1692–1763)

## Problem Set 1

1. Which of the following strings of brackets are balanced? In each case, explain carefully why the string is, or is not, balanced:

   (*i*)  $((()()$    (*ii*)  $(())()$    (*iii*)  $())(()$

2. Refer to the diagram for Problem 1.5 and suppose that there are $n$ railway wagons at $A$. The wagons can move from $A$ to $B$. It is assumed that the central track can accommodate all $n$ of them and that they travel only from left to right.

   (*i*) How many arrangements are possible at $B$, for $n = 1$, 2, 3, and 4?

   (*ii*) Suppose that there are five wagons at $A$, labelled 1, 2, 3, 4, 5. Is it possible to produce the arrangement 5, 1, 4, 2, 3 at $B$? What about 3, 2, 4, 1, 5?

3. Repeat the previous question but this time allow the wagons to move from $C$ to $A$ if necessary.

4. Consider a river system with $n$ sources which eventually merge to form a single stream. Assuming that no more than two streams merge at any point, we are interested in the number of ways that the mergers can take place.

   (i) Compile a table of values for $n = 1, 2, 3,$ and $4$ and then find (or guess) a general formula.

   (ii) If possible, find a connection with Problem 1.7.

5. In how many ways can a convex polygon with $n + 1$ sides (labelled $0, 1, 2, \ldots, n$) be divided into triangles by non-intersecting diagonals?

6. For $n \geq 0$, evenly distribute $2n$ points on the circumference of a circle. Let $a_n$ be the number of ways in which these $2n$ points can be paired off as $n$ chords where no two chords intersect.

   (i) Find $a_n$ for $n = 2, 3$ and then find (or guess) a general formula.

   (ii) If possible, find a connection with Problem 1.1.

7. Given two rows of boxes with $n$ boxes in each row:

   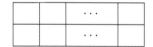

   In how many ways can you place the numbers $1, 2, \ldots, 2n$ in the boxes so that the numbers increase from left to right and so that each number in the bottom row is larger than the number in the box above it? Write down all the arrangements for $n = 1, 2, 3$ and $4$. Any conjectures?

*8. Given a strip of $n$ stamps, in how many ways can you fold them (at the perforations) into a *stack*? Do not distinguish between the front and back. For example, with $n = 2$ there is only one way to fold them, but for $n = 3$, there are two ways. Find the number of foldings for $n = 4$ and $n = 5$. You may find a strip of paper a useful tool. Any conjectures? (Now try $n = 6$.)

# 2

# Sets

SET theory is an essential ingredient of discrete mathematics and the language of set theory will be used throughout the rest of this book. You will see that it gives us a systematic and precise way to describe and solve many counting problems. In this chapter we review the fundamental notation and in later chapters we describe the connections between set theory and counting problems.

We shall take a *set* to be any collection of objects and we call these objects the *elements* of the set.

**Membership**  Let $A$ be a set. We write

$$x \in A$$

to mean that $x$ is an element of $A$ and we write

$$x \notin A$$

to mean that $x$ is *not* an element of $A$ (or, more simply, that $x$ is not in $A$).

EXAMPLE 2.1  We use $\mathbb{R}$ to denote the set of all real numbers, $\mathbb{Z}$ to denote the set of all integers and $\mathbb{N}$ to denote the set of all natural numbers (including 0). Then $x \in \mathbb{R}$ means that $x$ is a real number, $x \in \mathbb{Z}$ means that $x$ is an integer and $x \in \mathbb{N}$ means that $x$ is a natural number. Thus $\sqrt{2} \in \mathbb{R}$, but $\sqrt{2} \notin \mathbb{Z}$ and $-1 \in \mathbb{Z}$, but $-1 \notin \mathbb{N}$. ❑

**Set specification**  There are two useful ways to specify a set.

1. One way is simply to enclose the elements of the set in braces. For example

$$\{((())), (())(), ()(()), (())(), ()()()\}$$

is the set of all balanced strings of 3 left and 3 right brackets, considered in Example 1.2.

2. A second way to specify a set is to give a condition that the elements of the set must satisfy. For example, the set above can be described as

$$\{\, x \mid x \text{ is a balanced string of 3 pairs of brackets}\,\}.$$

EXAMPLE 2.2 The set of the first 10 natural numbers can be expressed as

$$\{0, 1, 2, 3, 4, 5, 6, 7, 8, 9\} \quad \text{or} \quad \{\, x \in \mathbb{N} \mid x \leq 9 \,\} \quad \text{or} \quad \{0, 1, \ldots, 9\}.$$

The set $E$ of all positive even integers can be written as

$$E = \{2,\ 4,\ 6,\ \ldots\,\} \quad \text{or} \quad E = \{\, 2k \mid k \in \mathbb{N} \text{ and } k > 0 \,\}. \quad \square$$

**The empty set** The set with no elements at all is called the *empty set* and we denote it by $\emptyset$ or $\{\ \}$.

EXAMPLE 2.3 The set

$$A = \{\, x \mid x \in \mathbb{R} \text{ and } x^2 + 1 = 0 \,\}$$

is the empty set. $\square$

**Subsets** Let $A$ and $B$ be two sets. Then $B$ is said to be a *subset* of $A$ if every element of $B$ is also an element of $A$ and we write this as $B \subseteq A$. We read $B \not\subseteq A$ as "$B$ is *not* a subset of $A$" — it means that there is an element of $B$ which does *not* belong to $A$.

*Note that the empty set is a subset of every set.*

EXAMPLE 2.4 Let

$$A = \{a,\ b,\ c,\ d\}, \quad B = \{b,\ c\} \quad \text{and} \quad C = \{b,\ c,\ e\}.$$

Then $B \subseteq A$, but $C \not\subseteq A$ (because $e \in C$ but $e \notin A$). $\square$

Sets themselves may be elements of another set.

EXAMPLE 2.5 Consider the set

$$A = \{a,\ b,\ c,\ \{a,\ b\},\ \{b,\ c\},\ 1,\ 2,\ \{3\}\,\}.$$

Then $\{a, b\} \in A$ and also $\{a, b\} \subseteq A$. On the other hand $3 \notin A$, $\{3\} \in A$, but $\{3\} \not\subseteq A$. Also $\{a, \{a, b\}\} \subseteq A$. The set $A$ has 8 elements. $\square$

**The power set**    The set of all subsets of a set $A$ is called the *power set* of $A$.

EXAMPLE 2.6    If $A = \{a, b\}$, then the subsets of $A$ are

$$\emptyset, \ \{a\}, \ \{b\}, \ \{a, b\},$$

and there are $2^2 = 4$ of them. Therefore, the power set of $A$ is

$$\{\emptyset, \ \{a\}, \ \{b\}, \ \{a, b\}\}.$$

If $A = \{a, b, c\}$, then the power set of $A$ is

$$\{\emptyset, \ \{a\}, \ \{b\}, \ \{c\}, \ \{a, b\}, \ \{a, c\}, \ \{b, c\}, \ \{a, b, c\}\},$$

and so $A$ has $2^3 = 8$ subsets. ❏

In Chapter 6 we shall see that if $A$ has $n$ elements, then there are $2^n$ subsets of $A$.

Here is a systematic way to list the subsets of a set. For example, to list the 16 subsets of $\{a, b, c, d\}$, proceed as follows:
- Begin with 16 empty pairs of braces { }.
- Put $a$ into the first 8 of these sets.
- Put $b$ into the first 4, skip 4, then put $b$ into the next 4.
- Put $c$ into the first 2, skip 2, put $c$ into the next 2, and so on.
- Put $d$ into the first set, skip 1, put $d$ into the next set, and so on.
- This process produces the 16 subsets of $\{a, b, c, d\}$. The last subset in the list is the empty set.

**Equality**    Two sets $A$ and $B$ are *equal* if they have the same elements.

EXAMPLE 2.7    We have $\{2, 3, 4\} = \{4, 2, 3\}$. Also $\{1, 1, 2\} = \{1, 2\}$ because the only elements in the set $\{1, 1, 2\}$ are 1 and 2. ❏

The best way to show that two sets $A$ and $B$ are equal is to show that $A \subseteq B$ and $B \subseteq A$. That is, show that every element of $A$ is an element of $B$ and show that every element of $B$ is an element of $A$. For an application, look ahead to Theorem 2.13.

**Union**  Let $A$ and $B$ be sets. The *union* of $A$ and $B$, denoted by $A \cup B$, is the set of elements that are in either $A$ *or* $B$. That is,

$$A \cup B = \{\, x \mid x \in A \text{ or } x \in B \,\}.$$

If $A_1, A_2, \ldots, A_n$ are sets, then the *union* of these sets is defined by

$$\bigcup_{k=1}^{n} A_k = \{\, x \mid x \in A_k \text{ for some } k,\ 1 \le k \le n \,\}.$$

EXAMPLE 2.8  Let $A = \{a, b, c, d, e\}$ and $B = \{1, 2, 3, 4, a, b\}$. Then
$$A \cup B = \{a, b, c, d, e, 1, 2, 3, 4\}. \quad \square$$

**Intersection**  Let $A$ and $B$ be sets. The *intersection* of $A$ and $B$, denoted by $A \cap B$, is the set of elements that are in both $A$ *and* $B$. That is,

$$A \cap B = \{\, x \mid x \in A \text{ and } x \in B \,\}.$$

If $A_1, A_2, \ldots, A_n$ are sets, then the *intersection* of these sets is defined by

$$\bigcap_{k=1}^{n} A_k = \{\, x \mid x \in A_k \text{ for all } k,\ 1 \le k \le n \,\}.$$

EXAMPLE 2.9  Let $A = \{a, b, c, d, e\}$ and $B = \{1, 2, 3, 4, a, b\}$. Then
$$A \cap B = \{a, b\}. \quad \square$$

**Complement**  Let $A$ and $B$ be sets. The set $A \setminus B$, called the *complement* of $B$ in $A$, is defined to be the set of elements that are in $A$ but *not* in $B$. That is,
$$A \setminus B = \{\, x \mid x \in A \text{ and } x \notin B \,\}.$$

EXAMPLE 2.10  Let $A = \{a, b, c, d, e\}$ and $B = \{1, 2, 3, 4, a, b\}$. Then

$$A \setminus B = \{c, d, e\}, \quad \text{and}$$
$$B \setminus A = \{1, 2, 3, 4\}. \quad \square$$

**Cardinality**  The number of different elements in the set $A$ is called the *cardinality* (or *size*) of $A$ and written $|A|$.

EXAMPLE 2.11  Let $A = \{a, b, c, d, e\}$ and $B = \{1, 2, 3, 4, a, b\}$. Then
$$|A| = 5, \quad |B| = 6,$$
$$|A \cup B| = 9, \quad |A \cap B| = 2,$$
$$|A \setminus B| = 3, \quad |B \setminus A| = 4. \quad \square$$

# The algebra of set theory

Ordinary algebra begins with the laws satisfied by numbers. For example, the *commutative laws*: $x + y = y + x$ and $xy = yx$; and the *associative laws*: $x + (y + z) = (x + y) + z$ and $x(yz) = (xy)z$. Similar laws hold for the operations of union and intersection of sets and because sets are different from numbers there are other laws which have no counterpart in the algebra of numbers. A sample of the laws of set theory is given by the following list of identities which hold for all sets $A$, $B$, $C$ and $X$.

(i)   $A \cup A = A$,  $A \cap A = A$

(ii)  $A \cup \emptyset = A$,  $A \cap \emptyset = \emptyset$

(iii) $A \cup B = B \cup A$,  $A \cap B = B \cap A$

(iv)  $A \cup (B \cup C) = (A \cup B) \cup C$,  $A \cap (B \cap C) = (A \cap B) \cap C$

(v)   $A \cup (B \cap C) = (A \cup B) \cap (A \cup C)$,  $A \cap (B \cup C) = (A \cap B) \cup (A \cap C)$

(vi)  $A \cup B \subseteq X$ if and only if $A \subseteq X$ and $B \subseteq X$

(vii) $X \subseteq A \cap B$ if and only if $X \subseteq A$ and $X \subseteq B$.

How do we know these identities are true? The answer is that we can *prove* them. Proof is central to mathematics and it is a good idea to get some practice as soon as possible. So we shall begin by proving (vi). As is customary in mathematics we set this out by stating what we wish to prove as a *theorem* followed by a *proof*.

*For more information about the logic used in proofs, see Chapter 11.*

In the following proof we shall use ideas from logic which will be dealt with in more detail in later chapters. Nevertheless, try to follow the reasoning as best you can. Later, after you've read Chapter 11 you can come back to this proof and see if your understanding has improved.

**Theorem 2.12** $A \cup B \subseteq X$ if and only if $A \subseteq X$ and $B \subseteq X$.

**Proof.** Suppose at first that $A \cup B \subseteq X$. We must show both that $A \subseteq X$ and that $B \subseteq X$. Now, if $x \in A$, then it is also the case that $x \in A \cup B$ and then $x \in X$ because $A \cup B \subseteq X$ by hypothesis. This means that every element of $A$ is also an element of $X$. Therefore, from the definition of subset, we have $A \subseteq X$.

*Try drawing a diagram to aid your understanding.*

Now we repeat the argument to show that $B \subseteq X$. That is, if $x \in B$, then it is also the case that $x \in A \cup B$ and then from our hypothesis we deduce that $x \in X$. This means that every element of $B$ is also an element of $X$, i.e., $B \subseteq X$. Thus we have shown that if $A \cup B \subseteq X$, then $A \subseteq X$ and $B \subseteq X$.

To complete the proof we must show that the converse holds. That is, we suppose that $A \subseteq X$ and $B \subseteq X$ and then show $A \cup B \subseteq X$. Thus our hypothesis in this case is $A \subseteq X$ and

$B \subseteq X$. If $x \in A \cup B$, then from the definition of $A \cup B$ there are two possibilities: either $x \in A$ or $x \in B$. If $x \in A$, then by hypothesis, $x \in X$. Similarly, if $x \in B$, then $x \in X$. Thus in both cases we have $x \in X$ and this shows that every element of $A \cup B$ is in $X$. Thus $A \cup B \subseteq X$. This completes the proof. ∎

When you first encounter this proof you may think that it is overly complicated. In a sense you would be right. After all, as soon as you begin to draw a diagram depicting the situation (as described in the next section) you will surely see that the result is true. Moreover, you may be able to shorten some of the arguments. For example, we have $A \subseteq A \cup B \subseteq X$ and therefore $A \subseteq X$, and so on. But the point of it all is to get you used to very precise arguments working directly from the given definitions.

Here is another example. This time we want to prove that two sets are equal. We do this by proving that each is contained in the other. The theorem is called the *distributive law* for union over intersection.

**Theorem 2.13** $A \cup (B \cap C) = (A \cup B) \cap (A \cup C)$.

**Proof.** By the definition of equality, we must show that

(2.14) $\qquad A \cup (B \cap C) \subseteq (A \cup B) \cap (A \cup C)$

and

(2.15) $\qquad (A \cup B) \cap (A \cup C) \subseteq A \cup (B \cap C)$.

To prove (2.14), we suppose that $x \in A \cup (B \cap C)$ and show that $x \in (A \cup B) \cap (A \cup C)$. If $x \in A \cup (B \cap C)$, then there are two cases: either $x \in A$ or else $x \in B \cap C$.

In the first case, i.e., $x \in A$, it is certainly true that $x \in A \cup B$ and that $x \in A \cup C$ because both $A \cup B$ and $A \cup C$ contain $A$. Thus in this case $x \in (A \cup B) \cap (A \cup C)$, as required.

In the second case we have $x \in B$ *and* $x \in C$. Thus again $x \in A \cup B$ and $x \in A \cup C$. Therefore, in this case as well, we have $x \in (A \cup B) \cap (A \cup C)$. Putting the two cases together we see that we have proved (2.14).

Conversely, to establish (2.15), we suppose that

$$x \in (A \cup B) \cap (A \cup C)$$

and show that from this it follows that $x \in A \cup (B \cap C)$.

Now, if $x \in (A \cup B) \cap (A \cup C)$, then $x \in A \cup B$ and $x \in A \cup C$. Thus we know that (*i*) $x \in A$ or $x \in B$, *and that* (*ii*) $x \in A$ or $x \in C$.

One possibility is that $x \in A$; in this case nothing more needs to be said. The other possibility is that $x \notin A$. In this case we see from $(i)$ that $x \in B$ and we see from $(ii)$ that $x \in C$; that is, $x \in B \cap C$. We conclude that either $x \in A$ or $x \in B \cap C$ and hence $x \in A \cup (B \cap C)$. This is true for all $x$ and so (2.15) has been established, as required.

This completes the proof. □

A careful look at this proof shows that it reduces the distributive law for $\cap$ and $\cup$ to the distributive law for the logical connectives "and" and "or". At this stage we take this property of "and" and "or" for granted, but after you have read Chapter 11 you will be able to verify it using a truth table.

## Venn diagrams

*John Venn (1834–1923)*

A Venn diagram is a pictorial representation of sets in the plane. In a Venn diagram, ovals (or other suitable shapes) are used to represent sets. In the following diagram, the shading represents $A \cap B$, $A \setminus B$, and $A \cup B$, respectively.

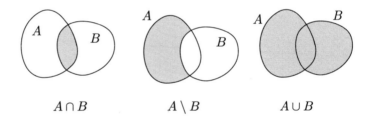

Figure 2.1
Basic Venn diagrams

Venn diagrams are most useful when dealing with a small number of sets. Here are some other examples involving three sets.

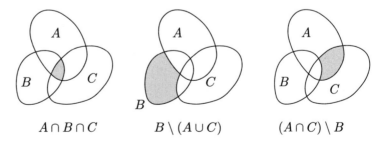

Figure 2.2
Venn diagrams with three sets

## De Morgan's laws

*Augustus De Morgan (1806–1871)*

De Morgan introduced his laws in connection with the "algebra of logic" — a subject which we shall study in Chapter 9. However, there is a close connection between logic and set theory and so his laws also apply to sets. We state the laws as a theorem, but leave the proof to you.

**Theorem 2.16** *If $A$ and $B$ are subsets of a set $X$, then*

(i) $\quad X \setminus (A \cup B) = (X \setminus A) \cap (X \setminus B)$
(ii) $\quad X \setminus (A \cap B) = (X \setminus A) \cup (X \setminus B)$.

As you can see, these laws deal with the complement operation for sets. Another, somewhat simpler law for the complement is the following.

**Theorem 2.17** *If $A$ is a subset of the set $X$, then $X \setminus (X \setminus A) = A$.*

This is illustrated by the following Venn diagram.

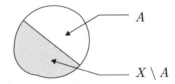

Figure 2.3
The complement of $A$ in $X$

## Summary

After reading this chapter you should know the meaning of

- set membership: $x \in A$;
- the empty set;
- subsets of a set;
- the power set of a set;
- union, intersection and complement of sets;
- the cardinality of a set;

*This is very important!* →
- proof;
- Venn diagrams.

# Problem Set 2

1. Use the notation of set theory to describe:
   - (*i*) The set of all odd integers between 2 and 10.
   - (*ii*) The set of all odd integers between 2 and 200.
   - (*iii*) The set of all odd integers.
   - (*iv*) The set of integers divisible by 4.

2. Which of the following statements are true?
   - (*i*) $\{2,4\} \subseteq \{1,2,3,4,5,6\}$.
   - (*ii*) $\{2\} \subseteq \{1,2,3,4,5,6\}$.
   - (*iii*) $2 \subseteq \{1,2,3,4,5,6\}$.
   - (*iv*) $2 \in \{1,2,3,4,5,6\}$.
   - (*v*) $\{2\} \in \{1,2,3,4,5,6\}$.
   
   Give reasons for your answers.

3. Write out the following sets, where $A = \{a, b, c, \{a, d\}\}$:
   - (*i*) $A \cup \{b, d, e\}$.
   - (*ii*) $A \cap \{b, d, e\}$.
   - (*iii*) $A \setminus \{a, b\}$.
   - (*iv*) $A \setminus \{c, d\}$.
   
   Then write down the sizes of each of the sets:

4. Write out a careful proof that $A \cap B = A$ implies $A \subseteq B$. Is the converse true? (First *write down* the converse of the implication.)

5. In Theorem 2.12 and Theorem 2.13 we proved two of the laws of set theory. Now prove the remaining laws of that section.

6. Prove both De Morgan's laws and draw diagrams to *illustrate* them. (Remember, a diagram can be an excellent guide to a proof, but it is not in itself a proof.)

7. Let $A$ be a set. Prove that $\{a\} \subseteq A$ if and only if $a \in A$.

*8. Consider Venn diagrams in which sets are represented by circles. Such a diagram divides the plane into a number of regions. How many regions can be obtained using three sets? How many for four sets? Is this enough to depict all possible relations between the sets?

# 3

# Functions

THIS chapter provides an essential link between the set theory discussed in Chapter 2 and the techniques of counting to be dealt with in the following chapters.

## What is a function?

*A function is a process.*

A function can be thought of as a process. This process, when applied to an element of a set, transforms it into an element of another set. That is, if $A$ and $B$ are sets, then a *function* or *mapping* $f : A \to B$ is a rule which assigns to each element $x$ in $A$ a *unique* element in $B$, denoted by $f(x)$. We say that $f$ is *defined* on $A$ or that $A$ is the *domain* of $f$. The *image* of $f$ is the set $\{\, f(x) \mid x \in A \,\}$.

*A function is a formula.*

EXAMPLE 3.1 The process which transforms each element $x \in \mathbb{Z}$ to its square is a function $f : \mathbb{Z} \to \mathbb{N}$ which we can write as $f(x) = x^2$. Sometimes it is convenient to write this function in the form
$$f : \mathbb{Z} \to \mathbb{N} : x \mapsto x^2. \quad \square$$

Not all functions can be described by such simple formulas. Indeed, in this book this is not always the most fruitful way to think of a function. Instead we can use some of the ideas of set theory to find other representations.

A case that comes up quite often in later chapters is that of a function $f$ defined on the set $\{1, 2, \ldots, n\}$ of the first $n$ positive integers. In this case the function is completely determined by the list of its *values*:

*A function is a list.*

$$(f(1), f(2), \ldots, f(n)).$$

In other words, a function of this type is nothing but an $n$-tuple, i.e., a sequence of values separated by commas and surrounded by a pair of brackets. We could also write this as $(x_1, x_2, \ldots, x_n)$, where $x_i = f(i)$ for $i = 1, 2, \ldots, n$.

EXAMPLE 3.2 If $f : \{1, 2, 3, 4, 5\} \to \{a, b, c\}$ is the function defined by $f(1) = f(3) = f(4) = b$ and $f(2) = f(5) = a$, then $f$ is completely described by the 5-tuple $(b, a, b, b, a)$. ❏

EXAMPLE 3.3 A balanced string of brackets, such as ()(()), can be thought of as a function from $\{1, 2, \ldots, 6\}$ to $\{\, (, ) \,\}$. If $f$ is the function corresponding to ()(()), then $f(1) = f(3) = f(4) = ($ and $f(2) = f(5) = f(6) = )$. This could get rather confusing because we are using brackets in two ways: as abstract symbols and also in the conventional mathematical way. However, we now see how to define a *string* of $n$ symbols: it is just the list of values of a function defined on the first $n$ positive integers. ❏

*A function is a set of ordered pairs.*

In practice not all functions will be defined on such convenient sets of integers. But taking our cue from these examples we see that all we really need to do to specify a function $f : A \to B$ completely is to give the set of all ordered pairs

$$\{\, (x, f(x)) \mid x \in A \,\}.$$

*A function is an arrow diagram.*

This way of defining a function leads to a very convenient pictorial description: the *arrow diagram*. We construct the diagram for $f : A \to B$ by drawing an arrow from $x$ to $f(x)$ for all $x$ in $A$.

EXAMPLE 3.4 Here is the arrow diagram for the function defined in Example 3.2.

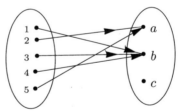

Figure 3.1
An arrow diagram.

In summary, a function can be thought of as a process, as a sequence, or as a set of ordered pairs. For $f$ to be a function from $A$ to $B$, the arrow diagram of $f$ must have exactly one arrow starting at each element of $A$. ❏

**One-to-one functions**

A function $f : A \to B$ is said to be *injective* or *one-to-one* if whenever $x \ne y$ in $A$, then $f(x) \ne f(y)$ in $B$. Equivalently, if whenever $f(x) = f(y)$ in $B$, then $x = y$ in $A$. This means that $f$ is one-to-one provided that, in its arrow diagram, no two arrows have the same end point.

EXAMPLE 3.5 The function $f : \{1,2,3\} \to \{a,b,c,d\}$ corresponding to the 3-tuple $(d,c,a)$ is one-to-one. □

## Onto functions

A function $f : A \to B$ is said to be *surjective* or *onto* if for any element $y \in B$, there is some element $x \in A$ such that $f(x) = y$. Equivalently, $f : A \to B$ is onto if and only if the image of $f$ equals $B$. Thus for $f$ to be onto, every element of $B$ in the arrow diagram of $f$ must be the end point of at least one arrow.

EXAMPLE 3.6 The following diagrams illustrate various types of functions.

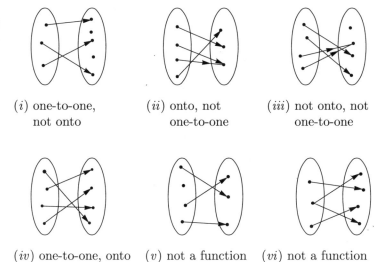

Figure 3.2
Examples of functions.

(i) one-to-one, not onto

(ii) onto, not one-to-one

(iii) not onto, not one-to-one

(iv) one-to-one, onto

(v) not a function

(vi) not a function

□

## The pigeonhole principle

Let $A$ and $B$ be finite sets. The examples above suggest the following facts:

(i)   If there is a one-to-one function $f : A \to B$, then $|A| \leq |B|$.

(ii)  If there is an onto function $f : A \to B$, then $|A| \geq |B|$.

(iii) If there is a one-to-one and onto function $f : A \to B$, then $|A| = |B|$.

Another way of stating (i) is the following

**Pigeonhole Principle.** *If $m$ objects are placed in $n$ boxes and if $m > n$, then at least one box contains two or more objects.*

If $O$ is the set of objects and if $B$ is the set of boxes, then assigning an object to a box describes a function $f : O \to B$. In this case, $f(x)$ is the box which contains $x$. If $m > n$, then this function cannot be one-to-one and therefore there exist objects $x$ and $y$ such that $f(x) = f(y)$. That is, $x$ and $y$ are in the same box.

**Problem 3.7** *If 5 numbers are chosen from* $\{1, 2, 3, 4, 5, 6, 7, 8\}$, *show that at least two of them sum to 9.*

**Solution.** To see this, we apply the pigeonhole principle by taking $b_1 = \{1, 8\}$, $b_2 = \{2, 7\}$, $b_3 = \{3, 6\}$ and $b_4 = \{4, 5\}$ to be the "boxes". If we choose 5 numbers, then some box must contain 2 of them and therefore those numbers sum to 9. ❑

**Remark**
Here is a more refined version of the pigeonhole principle:
*If $m$ objects are placed in $n$ boxes and if $m > kn$, then at least one box contains $k + 1$ objects.*

**Proof.** Suppose not. That is, suppose every box contains at most $k$ objects. Then the total number of objects is at most $kn$. This contradicts the assumption that there are more than $kn$ objects. Therefore, if there are more than $kn$ objects, at least one of the boxes contains at least $k + 1$ of them. ❑

## Bijections

If there is a one-to-one function $f$ from $A$ onto $B$, then we say that $A$ and $B$ are in *one-to-one correspondence*, or that there is a *one-to-one correspondence* between $A$ and $B$. The function $f$ is also called a *bijection* or a *bijective function*.

Suppose that $A$ and $B$ are *finite* sets. Then there is a one-to-one correspondence between $A$ and $B$ if and only if $|A| = |B|$. If $|A| = |B|$, then a function $f : A \to B$ is one-to-one if and only if it is onto. This is not true if $A$ and $B$ are infinite.

EXAMPLE 3.8 The function $f : \mathbb{N} \to \mathbb{N}$ defined by $f(x) = 2x$ is one-to-one but not onto. (Why?) ❑

## Permutations

When $A$ and $B$ are the *same set*, a one-to-one correspondence $f : B \to B$ is called a *permutation* of $B$.

Think of $B$ as a set of boxes and suppose that each box contains a single object. A permutation $f : B \to B$ describes how the *contents* of the boxes are rearranged. For box $x$, the object in $x$ is moved to the box $f(x)$.

A permutation $f$ of the set $\{1, 2, \ldots, n\}$ can also be pictured as an array with two rows:

$$\begin{array}{cccc} 1 & 2 & 3 & \ldots & n \\ f(1) & f(2) & f(3) & \ldots & f(n) \end{array}$$

Join each number $i$ in the top row to the corresponding number $i$ in the bottom row. This produces a diagram such as

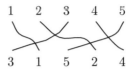

**Odd and even permutations**

The number of crossings depends on how the diagram is drawn but whether this number is even or odd does not. (The lines should go from top to bottom without doubling back and at most two lines should cross at any point.)

*This property of a permutation is called its* parity.

If the number of crossings is even, the permutation is said to be *even*; if it is odd, the permutation is said to be *odd*.

EXAMPLE 3.9 The permutation above has 4 crossings and therefore it is even. ☐

## Composition of functions

Given functions $f : A \to B$ and $g : B \to C$ it is possible to construct a new function $h : A \to C$ by first applying $f$ and then applying $g$. More precisely, for all $a \in A$, we define $h(a)$ by
$$h(a) = g(f(a)).$$
We write $h = g \circ f$ and call $h$ the *composition* of $f$ and $g$. The arrow diagram for $g \circ f$ is obtained by merging the arrow diagrams of $f$ and $g$. That is, the head of the arrow from $a$ to $f(a)$ joins the tail of the arrow from $f(a)$ to $g(f(a))$ and we think of the combination as a single arrow from $a$ to $g(f(a))$.

EXAMPLE 3.10 In this example, $A = \{a, b, c\}$, $B = \{1, 2, 3, 4\}$ and $C = \{x, y, z\}$. The functions $f$, $g$ and $g \circ f$ are defined by their arrow diagrams:

Figure 3.3
Composition
of functions

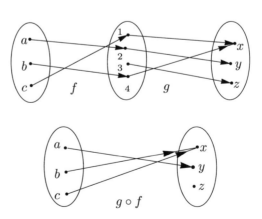

☐

# Brackets and railway wagons

We conclude this chapter by showing the connection between the bracketing and railway wagon problems of Chapter 1.

EXAMPLE 3.11 Let $\mathcal{A}$ be the set of all arrangements of four railway wagons that can be achieved beginning with the four wagons numbered 1 to 4 according to the scheme outlined in Problem 1.5. There are 14 arrangements; that is $|\mathcal{A}| = 14$.

Let $\mathcal{B}$ be the set of balanced strings of four pairs of brackets. In this case there are 14 ways to arrange the brackets (see Problem 1.3) and so $|\mathcal{B}| = 14$.

Since the sets $\mathcal{A}$ and $\mathcal{B}$ have the same size, we know that there must be a one-to-one correspondence between them. ☐

Our aim is to construct a one-to-one correspondence between the set $\mathcal{A}$ of arrangements of $n$ railway wagons and the set $\mathcal{B}$ of balanced strings of $n$ pairs of brackets.

This construction will make sense for any number of wagons, not just four, and it explains why the number of ways of arranging the railway wagons is always the same as the number of balanced strings of brackets.

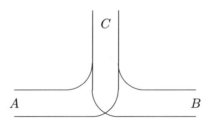

*Here is the construction.* Let "(" be regarded as an instruction to move a wagon from $A$ to $C$ in the diagram above and let ")" be regarded as an instruction to move a wagon from $C$ to $B$. We read the string of brackets from left to right when interpreting it as a set of instructions. Then, for example, beginning with 1234 at $A$, the instruction (()(())) produces the arrangement 4213 at $B$.

Applying a balanced string of brackets to the standard arrangement $1, 2, \ldots, n$ of wagons at $A$ produces an arrangement at $B$. In other words, we have a function $f$ from the set $\mathcal{B}$ of balanced strings of brackets to the set $\mathcal{A}$ of arrangements of railway wagons. For this to be a one-to-one correspondence we must check three things:

(i) That this makes sense. (That is, the function $f$ is *well-defined*.)

(ii) That different bracketings produce different arrangements at $B$. (That is, the function $f$ is *one-to-one*.)

(iii) That every arrangement at $B$ can be obtained by applying a balanced string of brackets to the standard arrangement at $A$. (That is, the function $f$ is *onto*.)

*f is well-defined*

Notice that in any balanced string of brackets there must be at least as many left brackets as right brackets in any initial segment of the string when reading from left to right. Thus, when interpreting brackets as instructions, whenever you come to a ")" there will always be a wagon at $C$ to move. This takes care of point $(i)$ above.

*f is one-to-one*

Now for $(ii)$. Suppose that $S_1$ and $S_2$ are two balanced strings of brackets which agree in their first $k$ positions but differ at the $(k+1)$-st position. In fact we may suppose that the $(k+1)$-st bracket of $S_1$ is "(" whereas the $(k+1)$-st bracket of $S_2$ is ")". (If not, we interchange $S_1$ and $S_2$.) If we now interpret $S_1$ and $S_2$ as instructions for moving wagons from $A$ to $C$ and from $C$ to $B$, then after the first $k$ instructions from either $S_1$ or $S_2$ we have exactly the same arrangements at $B$ and $C$. For example, if $S_1 = (()()())$ and $S_2 = (()())()$, then $k = 5$ and the situation after 5 moves is shown in the following diagram.

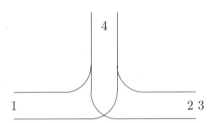

But the $(k+1)$-st instruction of $S_1$ is to move a wagon from $A$ to $C$ whereas the $(k+1)$-st instruction of $S_2$ is to move a wagon from $C$ to $B$. Thus the arrangements at $B$ and $C$ differ from this point on. This shows that $f(S_1) \neq f(S_2)$ and the claim made in $(ii)$ is true. (In the example $f(S_1) = 4123$ but $f(S_2) = 1423$.)

*f is onto*

Finally, suppose that we have an arrangement of $n$ wagons at $B$ which was produced by moving wagons from $A$ to $C$ and from $C$ to $B$ in some order. This process can be described by a string $S$ of $n$ left and $n$ right brackets. In fact $S$ must be a balanced string—the instruction to move wagon $n - k + 1$ from $A$ to $C$ corresponds to the $k$-th left bracket, and the instruction to move wagon $n - k + 1$ from $C$ to $B$ corresponds to its matching right bracket. This shows that every arrangement at $B$ can be obtained using a balanced string of brackets and hence the function $f$ is onto.

This proves that $f$ is a one-to-one correspondence. ❑

**Remarks**

1. This example shows that constructing a one-to-one correspondence between two sets is often far from easy!
2. It also illustrates how it is possible to know that two sets have the same size without knowing what that size is.
3. The full explanation of why the function is onto really requires the notion of *mathematical induction*. This is discussed in Chapter 13.

# Summary

After reading this chapter you should know about

- several ways to describe a function;
- arrow diagrams;
- one-to-one functions;
- onto functions;
- bijections;
- permutations;
- composition of functions.

# Problem Set 3

1. Define $f : \mathbb{N} \to \mathbb{N}$ by $f(x) = x + 1$. Determine whether or not $f$ is

   (a) one-to-one;   (b) onto.

2. Each of the following sets of pairs may or may not represent a function from $\{1, 2, 3\}$ to $\{a, b, c, d\}$.

   $\{(1,d),(2,b),(3,d)\}$  $\{(1,c),(2,a),(3,b)\}$  $\{(1,a),(3,b)\}$
   $\{(1,a),(1,c),(3,d)\}$  $\{(2,b),(3,c),(1,d)\}$

   (*i*) Identify the sets which represent functions and determine which of these are one-to-one.
   (*ii*) Explain clearly why each of the sets does or does not represent a function.
   (*iii*) Explain clearly why each of the sets does or does not represent a one-to-one function.

3. (i) Let $A = \{-1, 2, 3, 5, 7, 11\}$ and let $B = \{1, 2, \ldots, 200\}$. Is the function $f : A \to B$ given by $f(x) = x^2$ one-to-one?

   (ii) Now suppose that $A = \{-2, -1, 2, 3, 5, 7, 11\}$ and $B = \{1, 2, \ldots\}$. Is the function $f : A \to B$ given by $f(x) = x^2$ one-to-one?

4. Let $A = \{1, 2, 3, 4\}$, $B = \{a, b, c, d\}$ and $C = \{x, y, z, t\}$. Let $f : A \to B$ be defined by $f(1) = b$, $f(2) = c$, $f(3) = b$ and $f(4) = a$. Let $g : B \to C$ be defined by $g(a) = t$, $g(b) = x$, $g(c) = y$ and $g(d) = t$. Let $h : B \to A$ be defined by $h(a) = 2$, $h(b) = 1$, $h(c) = 1$ and $h(d) = 4$.

   (i) Find the composition $g \circ f$ of $f$ and $g$.

   (ii) Describe the functions $h \circ f$ and $f \circ h$.

5. (i) Let $A = \{1, 2\}$ and $B = \{a, b\}$. Write down all the one-to-one correspondences between $A$ and $B$.

   (ii) Let $A = \{1, 2, 3\}$ and $B = \{a, b, c\}$. Write down all the one-to-one correspondences between $A$ and $B$.

6. Given permutations $f$ and $g$ of $\{1, 2, \ldots, n\}$.

   (i) Show that $g \circ f$ is also a permutation of $\{1, 2, \ldots, n\}$.

   *(ii) Find out how the parity of $g \circ f$ depends on the parity of $f$ and $g$. Can you prove your conjectures?

7. Let $\mathcal{A}$ be the set of all ways of bracketing $x_0 x_1 x_2 x_3$ so that each pair of brackets contains two terms and let $\mathcal{B}$ be the set of all balanced strings of 3 left and 3 right brackets.

   (i) Write out the elements of $\mathcal{A}$ and $\mathcal{B}$.

   (ii) What are the cardinalities of $\mathcal{A}$ and $\mathcal{B}$?

   (iii) Define a function $f : \mathcal{A} \to \mathcal{B}$ by erasing all the $x_i$'s. For example, $f$ assigns $((x_0 x_1)(x_2 x_3))$ to $(()())$. Is $f$ one-to-one?

*8. Let $\mathcal{F}_n$ be the set of functions $f : \{1, 2, \ldots, n\} \to \{1, 2, \ldots, n\}$ such that

   (i) $f$ is increasing, i.e., if $i \leq j$, then $f(i) \leq f(j)$, and

   (ii) for all $i$, $f(i) \leq i$.

   Write down the sets $\mathcal{F}_n$ for $n = 1, 2, 3, 4$. Can you make any conjectures about $|\mathcal{F}_n|$?

# 4

# Counting Principles

IN this chapter we introduce two simple but basic rules of counting: the Addition Principle and the Multiplication Principle. These principles give set theoretic interpretations to the fundamental operations of arithmetic (addition and multiplication).

## The Addition Principle

The first principle relates addition to union of sets. The easiest case is when the sets have no elements in common.

**Disjoint union**  If the sets $A$ and $B$ have no elements in common, we say that they are *disjoint* and we write $A \cap B = \emptyset$. If $A$ and $B$ are disjoint sets, then

$$|A \cup B| = |A| + |B|.$$

More generally, if the sets $A_1, A_2, \ldots, A_n$ are *pairwise disjoint*, i.e., $A_i \cap A_j = \emptyset$ whenever $i \neq j$, then

$$|A_1 \cup A_2 \cup \cdots \cup A_n| = |A_1| + |A_2| + \cdots + |A_n|.$$

**Remarks**

1. When $A$ and $B$ are disjoint, an element of $A \cup B$ is in $A$ or $B$ but *not both*. When we use "or" to mean "in $A$ or in $B$, but not in both", it is called the *exclusive or*.

   *exclusive or*

2. If $A$ and $B$ overlap, then an element of $A \cup B$ could be in $A$ and $B$. We still say that the elements of $A \cup B$ are those that are in $A$ or $B$. This is the way the word "or" is generally used in mathematics — it is called the *inclusive or*.

   *inclusive or*

The combinatorial interpretation of disjoint union is:

**The Addition Principle.** *If one thing can be selected in $a$ ways and another thing can be selected in $b$ ways, then the number of different ways of selecting the first thing **or** the second thing is $a + b$.*

EXAMPLE 4.1 Suppose that there are 18 mathematics books and 11 physics books. Then the number of different ways of choosing just one book is $18 + 11 = 29$. ❏

**Union and intersection**

If the sets $A$ and $B$ are not disjoint, then we have:

$$|A \cup B| = |A| + |B| - |A \cap B|.$$

The reason is, in counting the sum $|A| + |B|$, we have counted the elements common to $A$ and $B$ twice and so we need to subtract $|A \cap B|$ from this sum to yield $|A \cup B|$.

EXAMPLE 4.2 Let $A = \{♠, ♡, ◇, ♣, 1, 2, \}$ and $B = \{1, 2, 3, 4\}$. Then

$$A \cup B = \{♠, ♡, ◇, ♣, 1, 2, 3, 4\} \quad \text{and} \quad B \cap A = \{1, 2\},$$

and so

$$|A| = 6, \ |B| = 4, \ |A \cap B| = 2, \ \text{and}$$
$$|A \cup B| = |A| + |B| - |A \cap B| = 8. \ ❏$$

We deal with the generalization to more than two sets in Chapter 7.

**Remark**
If $|A_1 \cup A_2 \cup \cdots \cup A_n| > m$, then at least one of the sets $A_1, A_2, \ldots, A_n$ has more than $\lfloor m/n \rfloor$ elements.
For if not, every one of the sets would have *at most* $\lfloor m/n \rfloor$ elements and as there are $n$ sets, their union would contain *at most* $m$ elements, which is not enough.
This is a form of the Pigeonhole Principle discussed in Chapter 3. The symbol $\lfloor m/n \rfloor$ means the largest integer $k$ such that $k \leq m/n$. For example, $\lfloor 7/3 \rfloor = 2$.

**The Multiplication Principle**

Even though this principle is quite simple it is the basis for most of the counting techniques in the next few chapters. We introduce it via the following problem.

**Problem 4.3** If 4 roads lead from $A$ to $B$ and 3 roads lead from $B$ to $C$, how many ways can you go from $A$ to $C$ via $B$?

**Solution.** Suppose that the roads from $A$ to $B$ are labelled 1, 2, 3 and 4 and that the roads from $B$ to $C$ are labelled $a, b$ and $c$. Then a journey from $A$ to $C$ can be described by a pair such as $(3, b)$. This means that road 3 is taken from $A$ to $B$, then road

$b$ is taken from $B$ to $C$. Thus the number of ways from $A$ to $C$ is the number of *ordered pairs* $(x, y)$, where $x \in \{1, 2, 3, 4\}$ and $y \in \{a, b, c\}$. In this case the number is easily seen to be 12. To confirm this you might list all the possibilities.

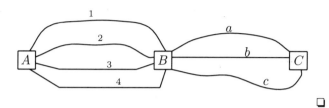

□

**Cartesian product**  In general, given sets $A$ and $B$, their *product* (also called their *Cartesian product*) is the set $A \times B$ of all *ordered pairs* $(x, y)$, where $x$ is an element of $A$ and $y$ is an element of $B$. We can write this as

$$A \times B = \{\, z \mid z = (x, y), \text{ where } x \in A \text{ and } y \in B \,\}.$$

The size of $A \times B$ is $|A| \times |B|$. This can be written

$$|A \times B| = |A| \times |B|.$$

EXAMPLE 4.4  Let $A = \{\spadesuit, \heartsuit, \diamondsuit, \clubsuit\}$ and $B = \{1, 2\}$. Then $A \times B$ is the set

$$\{(\spadesuit, 1), (\heartsuit, 1), (\diamondsuit, 1), (\clubsuit, 1), (\spadesuit, 2), (\heartsuit, 2), (\diamondsuit, 2), (\clubsuit, 2)\},$$

and $B \times A$ is the set

$$\{(1, \spadesuit), (1, \heartsuit), (1, \diamondsuit), (1, \clubsuit), (2, \spadesuit), (2, \heartsuit), (2, \diamondsuit), (2, \clubsuit)\}.$$

Notice that if $A$ and $B$ are distinct and non-empty, then $A \times B \neq B \times A$ but $|A \times B| = |B \times A|$.  □

EXAMPLE 4.5  Let $A$ be the set of letters from $a$ to $z$ and let $B$ be the set of digits 0 to 9. Then $A \times B$ is the set of all possible pairs of a letter followed by a digit. We can write

$$A \times B = \{\, (x,\ y) \mid x \text{ is a letter and } y \text{ is a digit} \,\}.$$

In this case $|A| = 26$ and $|B| = 10$, and so there are 260 pairs, i.e., $|A \times B| = 260$.  □

The formula for $|A \times B|$ suggests the following rule.

**The Multiplication Principle.** *If one thing can be selected in a ways and another thing can be selected in b ways, then the number of different ways of selecting the first **and** the second thing is ab.*

This principle actually goes beyond the formula for $|A \times B|$ because the set from which the second choice is made could depend on the first choice.

EXAMPLE 4.6 The number of ways of choosing a string of two *different* lower case letters is $26 \times 25 = 650$.

There are 26 ways of choosing the first letter of the required string. After choosing the first letter there are only 25 choices remaining for the second letter. (Note that the set from which these choices is made depends on the first letter.) ❑

It is also clear how to extend the principle to count selections of more than two things.

EXAMPLE 4.7 There are 9000 strings of four digits not beginning with 0.

Since the strings do not begin with 0, there are 9 ways to choose the first digit. For the second, third and fourth digit, there are 10 ways of choosing each of them. Hence the number of strings of four digits not beginning with 0, is $9 \times 10 \times 10 \times 10 = 9000$. ❑

**Remarks**

1. The definition of Cartesian product is easily extended from pairs to triples and so on. For example, let $A_1$, $A_2$ and $A_3$ be any 3 sets. Then

$$A_1 \times A_2 \times A_3 = \{\,(a_1, a_2, a_3) \mid a_1 \in A_1, a_2 \in A_2, a_3 \in A_3\,\}$$

   is the set of all triples from $A_1$, $A_2$ and $A_3$, in that order. Moreover, we have

$$|A_1 \times A_2 \times A_3| = |A_1|\,|A_2|\,|A_3|.$$

2. More generally, $A_1 \times A_2 \times \cdots \times A_n$ is the set of all $n$-tuples $(a_1, a_2, \ldots, a_n)$, where $a_i \in A_i$, i.e.,

$$A_1 \times A_2 \times \cdots \times A_n = \{\,(a_1, a_2, \ldots, a_n) \mid a_i \in A_i,\ i = 1, 2, \ldots, n\,\}.$$

   Its cardinality is

$$|A_1 \times A_2 \times \cdots \times A_n| = |A_1|\,|A_2|\,\ldots\,|A_n|.$$

EXAMPLE 4.8  The number of ways of choosing a string of 3 digits from the 10 digits $\{0, 1, 2, \ldots, 9\}$ is $10 \times 10 \times 10 = 1000$. ❑

EXAMPLE 4.9  The number of ways of choosing a string of 3 *different* digits from the 10 digits is $10 \times 9 \times 8 = 720$. ❑

Notice that in the two principles given in this chapter, addition corresponds to *exclusive or* and multiplication corresponds to *and then*.

EXAMPLE 4.10  The number of strings of at most three different lower case letters is

$$1 + 26 + 26 \times 25 + 26 \times 25 \times 24 = 16277.$$

*This example uses both the Addition and the Multiplication Principles.*

You can either choose no letters *or* you can choose one letter *or* you can choose one letter *and then* a different letter *or* you can choose one letter *and then* a different letter *and then* a letter different from the first two. ❑

**Remark**

*The formal definition of a function.*

Using Cartesian products we can give the precise definition of a function. That is, a function $f : A \to B$ is a *subset* of $A \times B$ with the property that for every element $x \in A$ there is exactly one element $y \in B$ such that $(x, y) \in f$. Actually, we've seen this before—the pairs $(x, y)$ correspond to the arrows in the arrow diagram for $f$ (as in Example 3.4).

# Theory of the Multiplication Principle

In this section we explain how the Multiplication Principle can be described in the language of set theory. What is perhaps surprising is that this apparently more general principle can be obtained from the Addition Principle.

To see this, suppose that we have a set $A$ and that for each element $x \in A$ we have a set $B_x$ which depends on $x$. We want to find the size of the set of ordered pairs

$$X = \{\, (x, y) \mid x \in A \text{ and } y \in B_x \,\}.$$

Notice that the sets $B_x$ can depend on $x \in A$ and therefore $X$ is generally not a Cartesian product (unless there is a set $B$ such that $B_x = B$ for all $x \in A$.)

On the other hand, for fixed $x \in A$, the ordered pairs $(x, y)$, where $y \in B_x$ form the Cartesian product $\{x\} \times B_x$. Therefore

$$X = \bigcup_{x \in A} (\{x\} \times B_x)$$

and furthermore, the sets $\{x\} \times B_x$ are disjoint. There is also a one-to-one correspondence between $\{x\} \times B_x$ and $B_x$ in which $(x, y)$ corresponds to $y$. Thus $|\{x\} \times B_x| = |B_x|$ and, from the Addition Principle, we have

(4.11) $$|X| = \sum_{x \in A} |B_x|.$$

In a great many of the examples that we deal with in this book the sets $B_x$ turn out to have the same size, say $b$. When this happens, we can write $|B_x| = b$, for all $x \in A$. If $|A| = a$, then formula (4.11) becomes
$$|X| = ab.$$
In fact, this is just the Multiplication Principle. The special case when all the $B_x$ are the same set corresponds to the formula for the size of a Cartesian product.

EXAMPLE 4.12 If $A = \{a, b, c, \ldots, z\}$ is the set of lower case letters, then to count the strings of two different lower case letters (as in Example 4.6) we may take $B_a = A \setminus \{a\}$, $B_b = A \setminus \{b\}$, ..., and $B_z = A \setminus \{z\}$. Then

$$|A| = 26 \quad \text{and} \quad |B_a| = |B_b| = \cdots = |B_z| = 25.$$

Therefore, in agreement with Example 4.6, there are $26 \times 25 = 650$ strings of two different lower case letters. ❑

# Summary

In this chapter you have learned the two basic rules of counting:

- the Addition Principle;
- the Multiplication Principle.

# Problem Set 4

1. (i) If $A = \{1, 2\}$ and $B = \{a, b, c\}$, write down the set $A \times B$.
   (ii) For $A = \{1, 2, 3, 4\}$, write down the subset of $A \times A$ consisting of all those ordered pairs $(x, y)$ such that $x \leq y$.
   (iii) For $A$ as above, let $D$ be the subset of $A \times A$ consisting of all ordered pairs $(x, y)$ such that $x = y$. What is $|D|$? Find a one-to-one correspondence between $D$ and $A$.

2. How many ways are there to form a string of 2 lower case letters?

3. How many strings of four digits are there if 0 is never used?

4. (i) How many strings of three upper case letters are there?
   (ii) How many strings of three upper case letters and digits are there?
   (iii) How many strings of three upper case letters and digits are there in which the first character is a letter?

5. If $A = \{a, b, c, d\}$ and $B = \{e, f\}$, write down the set $A \times B$. Then write down a one-to-one correspondence between the set $A \times B$ and the set $S$ of all subsets of $\{1, 2, 3\}$.

6. The number plates of a certain state consist of either three letters followed by three digits or else three digits followed by three letters. How many plates can be produced?

7. Given an alphabet of 20 consonants and 6 vowels.
   (i) In how many ways can we select a consonant and then a vowel?
   (ii) In how many ways can we make a two-letter string consisting of one consonant and one vowel?

8. In a town of 18,000 people everyone has three initials. Must there be two people with the same initials?

*9. Given finite sets $A$ and $B$, let $E$ be a subset of $A \times B$. For $a \in A$, let
$$E(a) = \{\, b \in B \mid (a, b) \in E \,\}$$
and for $b \in B$, let
$$E^{\vee}(b) = \{\, a \in A \mid (a, b) \in E \,\}.$$

Prove that
$$\sum_{a \in A} |E(a)| = \sum_{b \in B} |E^{\vee}(b)|.$$

# 5

# *Ordered Selections*

IN this chapter we use the *Multiplication Principle* to count the number of ways of arranging objects subject to various conditions.

**Arrangements with repetition**

Let us consider the following problems:

**Problem 5.1** *How many ways are there to place three apples in four boxes?*

**Problem 5.2** *How many ways are there to park three cars in four parking stations?*

**Problem 5.3** *How many strings of three symbols can be formed from the four symbols ♠, ♡, ◇, and ♣?*

**Problem 5.4** *How many functions are there from a set of size 3 to a set of size 4?*

Each of the above problems reduces to working out the number of ways to place the first object, *and then* the second object, etc. We use the Multiplication Principle to obtain the required number of ways of arranging all the objects.

For example, in Problem 5.1, there are 3 apples to be placed and for each apple there are 4 choices of a box in which to place it. Hence by the Multiplication Principle, there are $4 \times 4 \times 4 = 4^3$ arrangements.

All four problems are essentially the same and so the argument just given shows that in each case there are $4^3$ arrangements. In these arrangements, the order of the objects is important and repetition is allowed.

**The number of functions**

Let $A$ be a set of apples and let $B$ be a set of boxes. A particular arrangement of apples in boxes can be thought of as a *function* $f$ from $A$ to $B$. We write $f : A \to B$ and if $x \in A$ is an apple, we interpret $f(x)$ as the box containing it.

Now suppose we have $m$ apples and $n$ boxes. Suppose that the apples are labelled $1, 2, \ldots, m$ and that the boxes are labelled $1, 2, \ldots, n$. Then an arrangement of apples in boxes corresponds to a function $f : \{1, 2, \ldots, m\} \to \{1, 2, \ldots, n\}$. This arrangement is completely described by the $m$-tuple of numbers

$$\bigl(f(1), f(2), \ldots, f(m)\bigr).$$

That is, apple $i$ is placed in box $f(i)$. In constructing such an arrangement we have $n$ choices of box for each apple. Thus the Multiplication Principle tells us that there are

$$\underbrace{n \times n \times \cdots \times n}_{m \text{ factors}} = n^m$$

ways to place $m$ apples in $n$ boxes.

We can rephrase this in terms of set theory:

*There are $n^m$ functions from a set of size $m$ to a set of size $n$.*

### Remark

*Note that the empty function is one-to-one!*

There is exactly *one* function from the empty set to any other set; it is called the *empty function*. The complete explanation for this requires us to use the formal definition of a function as a subset of a Cartesian product.

## Arrangements without repetition

We modify the previous problems by requiring that the objects are selected without repetition:

**Problem 5.5** *How many arrangements of three apples in four boxes are there if we make the restriction that each box can have at most one apple in it?*

**Problem 5.6** *How many ways are there to park three cars in four parking spaces in a parking station?*

**Problem 5.7** *How many strings of three distinct symbols can be formed from the four symbols ♠, ♡, ♢, and ♣?*

**Problem 5.8** *How many one-to-one functions are there from a set of size 3 to a set of size 4?* (Recall from Chapter 3 that a function $f : A \to B$ is *one-to-one* if whenever $x \neq y$ in $A$, then $f(x) \neq f(y)$ in $B$.)

Again we use the Multiplication Principle. For example, in the apple problem, there are 4 ways to choose the first apple, but for the second apple there are only 3 choices and for the third apple only 2 choices. Thus the total number of arrangements is $4 \times 3 \times 2 = 24$.

The same argument applies to the other problems.

**The number of one-to-one functions**

Returning to the interpretation of these problems in terms of sets and functions we see that these ordered arrangements without repetition correspond to *one-to-one* functions.

Let us count the number of one-to-one functions $f : A \to B$ when $A = \{1, 2, \ldots, m\}$ and $|B| = n$.

There are $n$ choices for $f(1)$ but then only $n - 1$ choices for $f(2)$ because we are not permitted to have $f(1) = f(2)$. Similarly, there are $n - 2$ choices for $f(3)$ and so on until at last we have $n - (m - 1)$ choices for $f(m)$. The Multiplication Principle tells us that the number of ways to choose $f$ is

$$(5.9) \qquad n(n-1)(n-2)\ldots(n-m+1).$$

For $m \neq 0$ we define the *falling factorial* $n_{(m)}$ to be this number. That is,

*Some books use $^nP_m$ instead of $n_{(m)}$.*

$$n_{(m)} = n(n-1)(n-2)\ldots(n-m+1).$$

If $m = 0$, we have $n_{(0)} = 1$ because, when $m = 0$, the only function from $A$ to $B$ is the empty function and it is one-to-one.

We have shown that:

> There are $n_{(m)}$ one-to-one functions from a set of size $m$ to a set of size $n$.

If $|A| = m$, a one-to-one function $f \colon A \to B$ can be thought of as an *ordered arrangement* of $m$ distinct objects selected from the set $B$. Hence the total number of ordered arrangements of $m$ elements selected from $n$ elements, without repetition, is $n_{(m)}$.

EXAMPLE 5.10 The number of ways of choosing 5 students from 10 students and seating them in a row of 5 chairs is

$$10_{(5)} = 10 \times 9 \times 8 \times 7 \times 6 = 30240. \qquad \square$$

**The number of bijective functions**

If $A = \{1, 2, \ldots, m\}$ and $B$ also has $m$ elements, then a bijection $f : A \to B$ corresponds to an ordered arrangement

$$(f(1), f(2), \ldots, f(m))$$

of the elements of $B$. As $A$ and $B$ are finite and have the same size, a function $f : A \to B$ is a bijection if and only if it is a one-to-one function. Therefore the formula (5.9) above shows that the number of bijections from $A$ to $B$ is

*This number is called $m$ factorial and written $m!$. Note that $0! = 1$.*

$$m_{(m)} = m(m-1) \cdots 3 \cdot 2 \cdot 1.$$

For example, if $B = \{\star, \diamond, \bullet\}$, then the $3! = 6$ arrangements are

$$(\star, \diamond, \bullet), \ (\star, \bullet, \diamond), \ (\diamond, \star, \bullet), \ (\diamond, \bullet, \star), \ (\bullet, \star, \diamond) \ \text{and} \ (\bullet, \diamond, \star).$$

**The number of permutations**  If the sets $A$ and $B$ are the *same*, a one-to-one and onto function (i.e., a bijection) $f : B \to B$ is called a *permutation* of $B$. Hence:

There are $n!$ permutations of a set of size $n$.

## A combinatorial proof

In the last exercise of Problem Set 4 you were asked to prove a general counting formula. One way of proving it is to count a certain set "in two ways". Here we shall illustrate the method by giving a combinatorial proof of

*This theorem can also be proved directly from the definitions of $n!$ and $n_{(m)}$.*

**Theorem 5.11**
$$n_{(m)} = \frac{n!}{(n-m)!}.$$

**Proof.** We shall count the number of ways to place $n$ apples in $n$ boxes, with one apple in each box. We have just seen that there are $n!$ ways to do this. On the other hand each arrangement could be obtained by placing the first $m$ apples in the boxes (there are $n_{(m)}$ ways to do this) and then placing the remaining $n-m$ apples in the remaining $n-m$ boxes (there are $(n-m)!$ ways to do this). From the Multiplication Principle we must have $n! = n_{(m)}(n-m)!$ and the given formula follows from this. ❑

## Summary

After reading this chapter you should know the formulas for

- the number of functions from a set of $m$ elements to a set of $n$ elements;
- the number of one-to-one functions from a set of $m$ elements to a set of $n$ elements;
- the number of permutations of a set of size $n$.

## Problem Set 5

1. $(i)$  The 10 students in a certain tutorial group each hand in an assignment. These assignments are then given to 3 markers. In how many ways can this be done?

   $(ii)$  Express $(i)$ above in terms of sets.

**36** ORDERED SELECTIONS

2. A **byte** is a string of eight 0's and 1's. How many bytes are there?

3. How many ways are there to seat three students on four chairs in a row?

4. How many six digit numbers are there which do not repeat a digit and do not begin with 0?

5. ($i$) How many strings of length 3 start with 2 digits and end with one of the 26 capital letters of the alphabet?
   ($ii$) In how many ways can 500 students seat themselves in a room containing 550 seats? (Give your answer in terms of factorials.)

6. ($i$) How many four digit numbers greater than 1000 can be formed using the digits 0, 1, 2, 3 and 4?
   ($ii$) How many four digit numbers greater than 1000, with no repeated digit, can be formed using the digits 0, 1, 2, 3 and 4?

7. Four people are about to have a snack and there are eleven types of cake available. Each person chooses just one cake.
   ($i$) How many possibilities are there?
   ($ii$) How many possibilities are there if everyone has a different type of cake?

8. A restaurant has five entrées, seven main courses and ten desserts. In how many ways can you select two dishes on the condition that they must not both be from the same part of the menu?

9. ($i$) How many strings of 8 distinct letters can be made from the letters $\{a, b, c, d, e, f, g, h\}$?
   ($ii$) How many of the strings you found in ($i$) do not have any of the elements of $\{a, b, c\}$ next to each other?

*10. Let $C_n = \{\pm 1, \pm 2, \ldots, \pm n\}$. How many permutations $f : C_n \to C_n$ are there such that $f(-x) = -f(x)$ for all $x \in C_n$?

# 6

# Unordered Selections

UP until now we have considered *ordered* selections and we have found formulae for the number of selections when repetition is allowed and when it is not allowed. But this is only half the story. We also want to know how many ways there are to select things when order does *not* matter.

## Selections without repetition

Suppose we have 6 apples and we wish to choose 2 of them *without* regard to order. We can choose the first apple in 6 ways and the second apple in 5 ways. But each pair will have been chosen twice. So there are $30/2 = 15$ ways to choose a pair of apples without regard to order.

*Binomial coefficients*

We use the symbol
$$\binom{m}{k}$$
to represent the number of ways of choosing $k$ things, without repetition, from $m$ things. It is read "$m$ choose $k$" and called a *binomial coefficient* (for reasons which will become apparent in a moment).

*Some books use $^mC_k$.*

Let $A$ be a set of size $m$ and let $K = \{1, 2, \ldots, k\}$. Then a one-to-one function $f : K \to A$ describes an ordered selection of $k$ things (without repetition) from $A$: $f(1)$ is the first, $f(2)$ the second, and so on. From Chapter 5 we know that the number of ways of selecting $k$ (ordered) things from $A$ is $m_{(k)}$. We also know that there are $k!$ ways to arrange $k$ things and each arrangement corresponds to the *same* unordered selection. Thus the number of unordered selections is
$$\frac{m_{(k)}}{k!}.$$
That is, using formula (5.11), the number of ways to choose $k$ things from $m$ things is

(6.1) $$\binom{m}{k} = \frac{m_{(k)}}{k!} = \frac{m!}{k!\,(m-k)!}.$$

This is also the number of *subsets* of $A$ of size $k$ because such a subset is simply an unordered selection of $k$ elements of $A$.

EXAMPLE 6.2 The number of ways to select a committee of 4 from a group of 11 people is $\binom{11}{4} = 330$. ◻

## The binomial theorem

The binomial coefficients are so named because they appear in the following formula expressing the powers of $a + b$ (a *binomial*) in terms of the *monomials* $a^i b^j$.

**Theorem 6.3** *For any non-negative integer $m$, we have*

$$(a+b)^m = \sum_{k=0}^{m} \binom{m}{k} a^k b^{m-k}.$$

There are several combinatorial proofs of this formula. We shall give two of them.

**First Proof.** First observe that

$$(a+b)^m = \underbrace{(a+b)(a+b)\ldots(a+b)}_{m}$$

*The left-hand side of the binomial identity.*

and that when we multiply this out, each term of the expansion is obtained by choosing either an $a$ or a $b$ from each factor and then multiplying them together.

The term with exactly $k$ $a$'s is $a^k b^{m-k}$. The number of times this occurs in the expansion is just the number of ways to choose $k$ of the $m$ factors. But we know this is $\binom{m}{k}$. Thus the expansion is $\sum_{k=0}^{m} \binom{m}{k} a^k b^{m-k}$, as required. ◻

*The right-hand side of the binomial identity.*

**Second Proof.** This proof is more complicated than the first one but it shows how each part of the formula can be interpreted as counting certain functions.

We begin with disjoint sets $A$ (of size $a$) and $B$ (of size $b$) and we let $M$ be a set of size $m$. We first count the number of functions $f : M \to A \cup B$ from $M$ to $A \cup B$. Since the size of $A \cup B$ is $a+b$ and the size of $M$ is $m$, there are $(a+b)^m$ such functions.

*The left-hand side of the binomial identity.*

On the other hand we can count these functions by another method. Namely, we choose a subset $K$ of $M$ of size $k$, a function

$$g : K \to A$$

and a function

$$h : M \setminus K \to B.$$

These two functions $g$ and $h$ can be combined to produce a function
$$f : M \to A \cup B$$
by defining
$$f(x) = \begin{cases} g(x), & \text{if } x \in K, \\ h(x), & \text{if } x \in M \setminus K. \end{cases}$$
This produces all the functions from $M$ to $A \cup B$ which send exactly $k$ elements of $M$ to $A$. The set $K$ can be chosen in $\binom{m}{k}$ ways, $g$ can be chosen in $a^k$ ways, and $h$ can be chosen in $b^{m-k}$ ways and so by the Multiplication Principle there are
$$\binom{m}{k} a^k b^{m-k}$$
choices for these functions. Then, by the Addition Principle, to get *all* the functions $f : M \to A \cup B$, we must take the sum from $k = 0$ to $k = m$. That is, the number of functions $f : M \to A \cup B$ is

> The reason that we begin the summation from $k = 0$ is that there is exactly one function from the empty set to any other set: the empty function.

$$\sum_{k=0}^{m} \binom{m}{k} a^k b^{m-k}.$$

By equating this summation with the number of functions calculated above, we obtain the binomial theorem. ☐

EXAMPLE 6.4  Putting $a = 1$, $b = 2x$ and $m = 6$, we have
$$(1 + 2x)^6 = \sum_{k=0}^{6} \binom{6}{k} 1^k (2x)^{6-k}$$
$$= 64x^6 + 192x^5 + 240x^4 + 160x^3 + 60x^2 + 12x + 1.$$

# Binomial identities

- When we put $a = b = 1$ in the binomial theorem, we get the identity

(6.5) $$2^m = \sum_{k=0}^{m} \binom{m}{k}.$$

Note that the right-hand side of (6.5) is the number of subsets of a set of size $m$. On the other hand, to choose a subset we consider each element in turn and decide whether or not to include it in the subset. Thus for each element there are

*An alternating sum.*

- Putting $a = -1$ and $b = 1$ in Theorem 6.3, we obtain

$$(6.6) \qquad 0 = \sum_{k=0}^{m} (-1)^k \binom{m}{k},$$

provided $m \neq 0$. We shall see an application (and an interpretation) for this identity in Chapter 7.

- There is no way to choose $n$ things from $m$ things if $n > m$ and therefore

$$\binom{m}{n} = 0, \quad \text{if } n > m.$$

- A set of size $m$ has just one subset of size 0, namely the empty set, and therefore

$$\binom{m}{0} = 1.$$

- Notice that choosing a subset of size $k$ is equivalent to choosing its complement (of size $m - k$), and therefore we have

$$(6.7) \qquad \binom{m}{k} = \binom{m}{m-k}.$$

- A fundamental identity is

$$\binom{m+1}{k} = \binom{m}{k} + \binom{m}{k-1}.$$

*Blaise Pascal (1623–1662)*

This is the basis of "Pascal's triangle". That is, the binomial coefficients can be displayed in a triangular array such that each entry is the sum of the two numbers immediately above:

*Known in China as Yang Hui's triangle (ca. 1250).*

$$
\begin{array}{c}
1 \\
1 \quad 1 \\
1 \quad 2 \quad 1 \\
1 \quad 3 \quad 3 \quad 1 \\
1 \quad 4 \quad 6 \quad 4 \quad 1 \\
1 \quad 5 \quad 10 \quad 10 \quad 5 \quad 1 \\
\cdots
\end{array}
\qquad
\begin{array}{c}
\binom{0}{0} \\
\binom{1}{0} \quad \binom{1}{1} \\
\binom{2}{0} \quad \binom{2}{1} \quad \binom{2}{2} \\
\binom{3}{0} \quad \binom{3}{1} \quad \binom{3}{2} \quad \binom{3}{3} \\
\binom{4}{0} \quad \binom{4}{1} \quad \binom{4}{2} \quad \binom{4}{3} \quad \binom{4}{4} \\
\binom{5}{0} \quad \binom{5}{1} \quad \binom{5}{2} \quad \binom{5}{3} \quad \binom{5}{4} \quad \binom{5}{5} \\
\cdots
\end{array}
$$

2 choices and so, by the Multiplication Principle, there are $2^m$ ways to choose a subset. This is another way to see that identity (6.5) is true.

**Proof.** We count the subsets of size $k$ in the set $\{1, 2, \ldots, m+1\}$ in two ways. On the one hand, there are $\binom{m+1}{k}$ of these subsets. On the other hand, there are $\binom{m}{k}$ sets of size $k$ which do not contain $m+1$ (these are the subsets of size $k$ in $\{1, 2, \ldots, m\}$), and $\binom{m}{k-1}$ sets of size $k$ which do contain $m+1$ (first choose $m+1$, then choose a subset of size $k-1$ from $\{1, 2, \ldots, m\}$). Thus there are $\binom{m}{k} + \binom{m}{k-1}$ of these subsets altogether. Equating the results of these calculations establishes the identity. □

*Addition Principle!*

*A. Vandermonde (1735–1796)*

- The *Vandermonde identity* is

$$\binom{w+m}{n} = \sum_{k=0}^{n} \binom{w}{k}\binom{m}{n-k}.$$

*The identity was known to Zhu Shi-Jie (ca. 1303)*

**Proof.** To see this, count the number of ways of choosing $n$ people from a group of $w$ women and $m$ men. On the one hand there are $\binom{w+m}{n}$ such subsets. This is the left hand side of the formula. On the other hand we could first choose $k$ women (in $\binom{w}{k}$ ways) and then complete the set of $n$ people by choosing $n-k$ men (in $\binom{m}{n-k}$ ways). Thus there are $\binom{w}{k}\binom{m}{n-k}$ ways to choose $k$ women and $n-k$ men. To get all possible subsets we must add these values for $k = 0, 1, \ldots, n$. This gives the right hand side of the formula. □

*Multiplication Principle!*

## Selections with repetition

The binomial coefficient $\binom{n}{m}$ gives the number of ways to select $m$ things from $n$ things without repetition. What is the number of selections if *repetition* is allowed?

Instead of going directly to the answer to this question we begin with a slightly different problem.

**Problem 6.8** *We have a supply of flowers of four different colours: red, white, yellow and blue. How many ways are there to choose six flowers, provided we choose at least one flower of each colour?*

In tackling this problem we shall suppose that there are at least six flowers available of each colour. Also, we do not distinguish between individual flowers — all we care about are the colours. For example, we could choose 3 red and 3 white, or 2 red, 2 white, a yellow and a blue, etc.

**Solution.** We begin with a row of six vases. Then, given a selection of flowers, we place them in the vases, with the red ones first, followed by the white, followed by the yellow, followed by the blue. Next we draw a vertical line between the vases where the colours change. There are five spaces between the vases and only three places where the colours change. Thus we will have drawn a line in three of the five places.

Conversely, each selection of flowers corresponds to a selection of three places in which to draw the separating lines: the red flowers placed to the left of the first line, the white flowers between the first and second line, and so on.

Therefore the number of ways to choose six flowers, allowing repetition of colours and choosing at least one flower of each colour is the number of ways to choose 3 things from 5 things, namely $\binom{5}{3} = 10$. ❏

The method just outlined applies in general. That is, suppose we want to choose $m$ things from $n$ things, allowing repetition, but insist on choosing at least one of each thing. To obtain such a selection we begin with a row of $m$ boxes. We then choose $n-1$ of the $m-1$ spaces between the boxes and put a divider at each chosen place. We put a copy of the first thing in each of the boxes to the left of the first divider, then a copy of the second thing in the boxes between the first and second dividers, and so on. We see from this that the total number of selections is the number of ways to choose dividers, namely

$$\binom{m-1}{n-1}.$$

*This is the number of ways to choose $m$ things from $n$ things, with at least one thing of each kind, and allowing repetition.*

Now we can return to our original question:

**Problem 6.9** *In how many ways can we select $m$ things from $n$ things if repetition is allowed?*

**Solution.** To answer this question we use a common mathematical device—we reduce it to a previously solved problem. In this case, we see that we can obtain our selections by first choosing $m + n$ things from $n$ things (allowing repetition and choosing at least one thing of each type) and then throwing away one thing of each type. This leaves us with a selection of $m$ of the $n$ things.

We have just seen that the number of ways to select $m+n$ things from $n$ things (allowing repetition and choosing at least one thing of each type), is $\binom{m+n-1}{n-1}$ and so this must be the answer to our original question. □

From identity (6.7) above we see that $\binom{m+n-1}{n-1} = \binom{m+n-1}{m}$ and therefore:

*The number of ways to select $m$ things from $n$ things, allowing repetition, but without regard for order, is*
$$\binom{m+n-1}{m}$$

EXAMPLE 6.10 From the formula just given, the number of ways to choose 6 flowers from a supply of flowers of 4 different colours is $\binom{6+4-1}{6} = 84$.

We can also solve this directly as follows. First we draw 6 crosses to represent a choice of 6 flowers.

$$\times \quad \times \quad \times \quad \times \quad \times \quad \times$$

*This method applies in general and is another way to think about selections with repetition.*

We can show which crosses represent flowers of different colours by using 3 vertical lines to divide the crosses into 4 groups. Then, we replace each cross to the left of the first vertical line by flowers of the first colour, each cross between the first and second line by flowers of the second colour and so on until the crosses to the right of the last line are replaced by flowers of the fourth colour. (Note that there may be no crosses between some of the lines.) For example, the following is a possible way to add 3 vertical lines

$$\times \quad \times \quad \times \quad | \quad | \quad \times \quad | \quad \times \quad \times$$

The question now reduces to finding the number of ways to add the 3 vertical lines. This can be done as follows. We add 3 more crosses to the 6 crosses above to get 9 crosses. Then to get a possible arrangement, we change 3 of these 9 crosses into vertical lines. Thus the number of possible outcomes is the number of ways of choosing 3 crosses from 9 crosses and it is given by

$$\binom{9}{3} = 84,$$

as before. □

EXAMPLE 6.11 The number of ways to choose four letters from the set $\{A, B, C\}$, allowing repetition, but without regard for order, is $\binom{4+3-1}{4} = 15$. The arrangements are:

$$
\begin{array}{ccccc}
AAAA & AAAB & AAAC & AABB & AABC \\
BBBB & ABBB & BBBC & AACC & ABBC \\
CCCC & ACCC & BCCC & BBCC & ABCC
\end{array}
$$

## Balls in boxes

There is also an interpretation of "unordered selection with repetition" in terms of placing balls in boxes:

**Problem 6.12** *Suppose that we have $m$ indistinguishable balls and $n$ boxes. How many arrangements of balls in boxes are there?*

**Solution.** Placing a ball in a box can be regarded as *selecting* that box, and so we are just asking for the number of ways of selecting $m$ things from $n$ things, allowing repetition. Thus the number of ways of placing $m$ indistinguishable balls in $n$ boxes is

$$\binom{m+n-1}{m}.$$

This is also the number of solutions to

$$x_1 + x_2 + \cdots + x_n = m,$$

where for each $i$, $x_i$ is a non-negative integer. Think of box $i$ containing $x_i$ balls. ☐

# Summary

The formulas that you have learnt in this and the previous chapter can be summarized in the following table showing the number of ways to select $m$ things from $n$ things.

| selection | ordered | unordered |
|---|---|---|
| with repetition | $n^m$ | $\binom{m+n-1}{m}$ |
| without repetition | $n_{(m)}$ | $\binom{n}{m}$ |

# Problem Set 6

1. Write down the expansion of
   - (i) $(x+2y)^5$,
   - (ii) $(2x-y)^6$.

2. You have a deck of fifty-two cards.
   - (i) How many ways are there of choosing a hand of five cards?
   - (ii) How many of them contain the queen of hearts?
   - (iii) In how many ways can four hands of five cards each be given to four players?
   - (iv) In how many ways can four hands of five cards be selected from the deck?

3. Consider the set $\{a,b,c,d,e,f\}$. How many ways are there of choosing four letters from this set
   - (i) if no letter is chosen twice?
   - (ii) if repetitions are allowed?

4. (i) In how many ways can thirteen cards be chosen from a deck of fifty-two cards?
   (ii) In how many ways can fifty-two cards be divided into four lots of thirteen?

5. How many different outcomes are possible if seven identical dice are thrown? (An *outcome* is the collection of numbers, with repetition, visible on the top faces of the dice.)

6. Given a large supply of jelly beans of 10 different colours, how many ways are there to make up a bag of 5 jellybeans?

*7. Find a formula for the number of solutions to

$$x_1 + x_2 + \cdots + x_n < m,$$

*Try some small examples first.*

where for each $i$, $x_i$ is a non-negative integer.

*8. Given $n$, how many sequences $(k_1, k_2, \ldots, k_r)$ of positive integers are there whose sum is $n$? For example, when $n$ is 4, the sequences are $(1,1,1,1)$, $(1,1,2)$, $(1,2,1)$, $(2,1,1)$, $(2,2)$, $(1,3)$, $(3,1)$ and $(4)$.

# 7

# The Inclusion-Exclusion Principle

THE main result of this chapter is a formula for the size of the union of sets. We first review the situation for unions of two and three sets before moving on to the general formula.

**The union of two sets**

Recall that in Chapter 4, we gave the formula

$$|A \cup B| = |A| + |B| - |A \cap B|,$$

for the size of the union of the sets $A$ and $B$.

This came about as follows. If we count $A \cup B$ by adding the size of $A$ to the size of $B$, each of the elements of $A \cap B$ has been counted twice. To compensate for this we must subtract $|A \cap B|$ from $|A| + |B|$ to get $|A \cup B|$.

Next, if $A$ is a subset of $X$, then

$$|X \setminus A| = |X| - |A|.$$

If $B$ is also a subset of $X$, then

$$|X \setminus (A \cup B)| = |X| - |A| - |B| + |A \cap B|.$$

EXAMPLE 7.1 In a class of 320, there are 198 computer science students and 130 mathematics students and 108 of these people are taking both computer science and mathematics.

(*i*) How many students are taking either computer science or mathematics?

(*ii*) How many are taking neither computer science nor mathematics?

**Solution.** Let $X$ be the set of all people in the class, $A$ the set of all students taking computer science and $B$ the set of all students taking mathematics. Then

$$|A| = 198, \ |B| = 130 \ \text{and} \ |A \cap B| = 108,$$

so that $|A \cup B| = |A| + |B| - |A \cap B| = 198 + 130 - 108 = 220$. Hence there are 220 students taking either computer science or mathematics.

Moreover $|X| = 320$ and therefore

$$|X \setminus (A \cup B)| = 320 - 220 = 100.$$

Thus there are 100 students taking neither computer science nor mathematics. ❏

## The union of three sets

Now we extend the formula for the size of a union to three sets $A$, $B$ and $C$.

If we try to count $A \cup B \cup C$ by adding (i.e., *including*) $|A|$, $|B|$ and $|C|$ we will have counted the elements in $A \cap B$, $A \cap C$ and $B \cap C$ twice. Therefore we should *exclude* these elements, that is, subtract $|A \cap B|$, $|A \cap C|$ and $|B \cap C|$ to compensate. But then the elements of $A \cap B \cap C$ will have been excluded once too often and so we must *include* them again.

This means that for three sets, we have the *Inclusion-Exclusion Principle*:

$$|A \cup B \cup C| = |A| + |B| + |C| - |A \cap B| - |A \cap C| - |B \cap C| + |A \cap B \cap C|.$$

EXAMPLE 7.2 Suppose in a tutorial of 10 people 3 have prepared solutions, 5 have read the text book, 5 have read the lecture notes, 2 have prepared solutions and read the book, 2 have prepared solutions and read the notes, 2 have read the book and read the notes and 1 has done all three. How many have done none of these things?

**Solution.** Let $A$ be the set of those people who have prepared solutions, $B$ the set of those who have read the text book and $C$ the set of those who have read the lecture notes. Then

$$|A| = 3, |B| = 5, |C| = 5, |A \cap B| = |A \cap C| = |B \cap C| = 2$$
$$\text{and } |A \cap B \cap C| = 1.$$

Now count how many have done at least one of the things. By the Inclusion-Exclusion Principle this is equal to

$$|A \cup B \cup C| = |A| + |B| + |C|$$
$$- |A \cap B| - |A \cap C| - |B \cap C| + |A \cap B \cap C|$$
$$= 3 + 5 + 5 - 2 - 2 - 2 + 1 = 8.$$

Therefore only 2 have done none of the things. ❏

EXAMPLE 7.3 In each case below we shall attempt to compute $|A \cup B \cup C|$ from the given information.

(i) $|A| = 5$, $|B| = 12$, $|C| = 7$, $|A \cap B| = 2$, $|A \cap C| = 4$, $|B \cap C| = 3$ and $|A \cap B \cap C| = 2$.

(ii) $|A| = 21$, $|B| = 31$, $|C| = 7$, $|A \cap B| = 3$, $|A \cap C| = 4$, $|B \cap C| = 13$ and $|A \cap B \cap C| = 2$.

**Solution.** Here is a complete Venn diagram for (i):

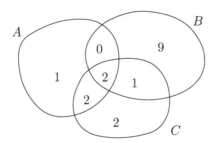

In this case it is possible to compute $|A \cup B \cup C|$ directly from the diagram or from the formula

$$|A \cup B \cup C| = |A| + |B| + |C|$$
$$- |A \cap B| - |A \cap C| - |B \cap C| + |A \cap B \cap C|$$
$$= 5 + 12 + 7 - 2 - 4 - 3 + 2 = 17.$$

In (ii), since $|C| = 7$, $|B \cap C|$ is at most equal to 7. But $|B \cap C|$ is given to be 13. This contradiction shows that the sets $A, B, C$ as described in (ii) cannot exist. Thus it does not make sense to compute $|A \cup B \cup C|$. ❑

EXAMPLE 7.4 How many numbers between 1 and 200 (inclusive) are divisible by at least one of the numbers 5, 7 or 11?

**Solution.** Let $A$ be the set of all numbers between 1 and 200 that are divisible by 5, let $B$ be the set of all numbers between 1 and 200 that are divisible by 7, and let $C$ be the set of all numbers between 1 and 200 that are divisible by 11. Then

$$|A| = 40, \ |B| = 28 \quad \text{and} \quad |C| = 18.$$

Now $|A \cap B|$ is the set of all numbers between 1 and 200 that are divisible by 5 and 7, i.e., by 35, and so

$$|A \cap B| = 5.$$

Similarly $|A \cap C|$ is the set of all numbers between 1 and 200 that are divisible by 5 and 11, i.e., by 55, and so

$$|A \cap C| = 3.$$

And $|B \cap C|$ is the set of all numbers between 1 and 200 that are divisible by 7 and 11, i.e., by 77, and so

$$|B \cap C| = 2.$$

Finally $|A \cap B \cap C|$ is the set of all numbers between 1 and 200 that are divisible by 5, 7 and 11, i.e., by 385, and so

$$|A \cap B \cap C| = 0.$$

There are $|A \cup B \cup C|$ numbers between 1 and 200 that are divisible by at least one of 5, 7 or 11 and by the Inclusion-Exclusion Principle, $|A \cup B \cup C|$ equals

$$|A| + |B| + |C| - |A \cap B| - |A \cap C| - |B \cap C| + |A \cap B \cap C| = 76.$$

❏

## The general inclusion-exclusion principle

Let $A_1, A_2, \ldots, A_n$ be $n$ sets. Then

$$|A_1 \cup A_2 \cup \cdots \cup A_n| = \sum_i |A_i| - \sum_{i<j} |A_i \cap A_j| + \ldots$$
$$+ (-1)^{n-1} |A_1 \cap A_2 \cap \cdots \cap A_n|.$$

Here is a sketch of why this is true.

Suppose that $x \in A_1 \cup A_2 \cup \cdots \cup A_n$ is in *exactly* $m$ of the sets $A_1, A_2, \ldots, A_n$. How many of the sets of the form $A_{i_1} \cap A_{i_2} \cap \cdots \cap A_{i_k}$ is $x$ in? The answer is $\binom{m}{k}$ because it is just the number of ways of choosing $k$ of the sets which contain $x$.

This means that on the right hand side of the formula $x$ will be counted

$$\binom{m}{1} - \binom{m}{2} + \cdots + (-1)^{m-1} \binom{m}{m}$$

times.

However, from equation (6.6) of the previous chapter, we know that for $m \neq 0$,

$$\binom{m}{0} - \binom{m}{1} + \cdots + (-1)^m \binom{m}{m} = 0,$$

and so

$$\binom{m}{1} - \binom{m}{2} + \cdots + (-1)^{m-1} \binom{m}{m} = \binom{m}{0} = 1.$$

Therefore the right hand side of the formula actually counts $x$ exactly once.

# 50 THE INCLUSION-EXCLUSION PRINCIPLE

*The perceptive reader will see that we are counting onto functions!*

EXAMPLE 7.5 You have 5 different apples and 4 different boxes. In how many ways can you arrange the apples in the boxes so that each box contains at least one apple?

**Solution.** Let $X$ be the set of all possible arrangements of the apples in the boxes and let $A_i$ be the set of arrangements for which the $i$-th box is empty. Then

$$|X| = 4^5 = 1024,$$
$$|A_i| = 3^5 = 243, \quad (i = 1, 2, 3, 4)$$
$$|A_i \cap A_j| = 2^5 = 32, \quad (i, j = 1, 2, 3, 4), \ i \neq j$$
$$|A_i \cap A_j \cap A_k| = 1, \quad (i, j, k = 1, 2, 3, 4), \ i, j, k \text{ all different}$$
$$|A_1 \cap A_2 \cap A_3 \cap A_4| = 0.$$

Thus the total number of arrangements with at least one box empty is

$$|A_1 \cup A_2 \cup A_3 \cup A_4|$$
$$= \sum_{i=1}^{4} |A_i| - \sum_{i<j} |A_i \cap A_j|$$
$$+ \sum_{i<j<k} |A_i \cap A_j \cap A_k| - |A_1 \cap A_2 \cap A_3 \cap A_4|$$
$$= 4 \times 243 - 6 \times 32 + 4 \times 1 - 0 = 784.$$

The number of sets of the form $A_i \cap A_j$ is $\binom{4}{2}$ and the number of sets of the form $A_i \cap A_j \cap A_k$ is $\binom{4}{3}$, therefore the number of ways to arrange the apples in the boxes so that each box contains at least one apple is $1024 - 784 = 240$. ❑

If all the sets $A_i$ are subsets of a set $X$, the number of elements of $X$ **not** in any of the $A_i$ is

$$|X \setminus (A_1 \cup A_2 \cup \cdots \cup A_n)| = |X| - \sum_i |A_i| + \sum_{i<j} |A_i \cap A_j| - \cdots$$
$$- (-1)^{n-1} |A_1 \cap A_2 \cap \cdots \cap A_n|.$$

*Quite a difficult problem!*

EXAMPLE 7.6 An Engineering student is given 5 glasses, each containing a different type of beer. He is told the names of the beers but not which glasses they are in. After drinking each beer he names it. In how many ways can he get every answer wrong?

**Solution.** Let $X$ be the set of all possible guesses. A guess is just a permutation of the 5 names and so $|X| = 5! = 120$.

Let $A_1$ be the set of guesses in which glass 1 is correctly named, let $A_2$ be the set of guesses in which glass 2 is correctly named, and so on. Then for the sets $A_1, \ldots, A_5$ we have

$$|A_i| = 4! = 24 \quad \text{for every } i,$$
$$|A_i \cap A_j| = 3! = 6 \quad \text{for } i < j,$$
$$|A_i \cap A_j \cap A_k| = 2! = 2 \quad \text{for } i < j < k,$$
$$|A_i \cap A_j \cap A_k \cap A_\ell| = 1 \quad \text{for } i < j < k < \ell,$$
$$|A_1 \cap A_2 \cap A_3 \cap A_4 \cap A_5| = 1.$$

Therefore, the number of guesses which have no beer correctly named is

$$120 - 5 \times 24 + 10 \times 6 - 10 \times 2 + 5 \times 1 - 1 = 44. \quad \square$$

# Summary

After reading this chapter you should be able to use the Inclusion-Exclusion Principle to

- calculate the size of the union of several sets;
- calculate the size of the complement of the union of several sets.

# Problem Set 7

1. If possible, compute $|A \cup B \cup C|$ from the given information. If it is not possible, explain why.

    (i) $|A| = 12$, $|B| = 13$, $|C| = 7$, $|A \cap B| = 5$, $|A \cap C| = 4$, $|B \cap C| = 3$ and $|A \cap B \cap C| = 2$.

    (ii) $|A| = 12$, $|B| = 13$, $|C| = 7$, $|A \cap B| = 10$, $|A \cap C| = 7$, $|B \cap C| = 13$ and $|A \cap B \cap C| = 2$.

    (iii) $|A| = 12$, $|B| = 13$, $|C| = 7$, $|A \cap B| = 10$, $|A \cap C| = 4$, $|B \cap C| = 6$ and $|A \cap B \cap C| = 2$.

2. You have 6 different apples and 3 different boxes. In how many ways can you arrange the apples in the boxes so that each box contains at least one apple?

3. In a group of 400 students, 300 are doing mathematics, 250 are doing physics and 200 are doing chemistry. Furthermore, 210 are doing mathematics and physics, 120 are doing mathematics and chemistry and 80 are doing physics and chemistry. Only 40 are doing all three subjects. How many students are not doing any of these subjects?

4. In a group of 50 participants at a recent international meeting

   30  speak English
   18  speak German
   26  speak French
   9   speak both English and German
   16  speak both English and French
   8   speak both French and German
   47  speak at least one of English, French or German

   (i) How many people in the group cannot speak English, French or German?
   (ii) How many people in the group can speak all three languages?

5. How many numbers between 1 and 100 (inclusive) are divisible by at least one of the numbers 3, 5 or 7?

6. Let $X$ be the set of all 5-element subsets of $\{1, 2, \ldots, 50\}$. How many elements of $X$ contain at least one element of $\{1, 2, 3, 4, 5\}$?

7. (i) Given $A = \{1, 2, 3, 4\}$. How many permutations $f : A \to A$ have the property that $f(i) = i$ for at least one value of $i$?
   (ii) Repeat (i) for $A = \{1, 2, 3, 4, 5\}$.

8. How many ways can you place 6 indistinguishable apples in 4 boxes with at most 3 apples in each box?

*9. How many permutations $f : \{1, 2, 3, 4, 5\} \to \{1, 2, 3, 4, 5\}$ have the property that $f(i) = i$ for at least two values of $i$?

*10. Use the Inclusion-Exclusion Principle to find a formula for the number of onto functions from a set of size $m$ to a set of size $n$. (cf. Example 7.5.)

# 8

# Multinomial Coefficients

IN this chapter we study a simple extension of the binomial theorem which will enable us to count permutations involving indistinguishable objects.

**Permutations with repetition**

To illustrate the main idea, we begin with the following

**Problem 8.1** *How many ways can we permute the letters of*

$$WOOLLOOMOOLOO$$

*to form different "words"? (The "words" don't have to make sense.)*

The answer is not 13! because, for example, simply permuting the $O$'s among themselves will not change the word.

**Solution.** Observe that there are 3 $L$'s, 1 $M$, 8 $O$'s and 1 $W$. To count the number of arrangements imagine that we have 13 boxes:

We shall place the letters of $WOOLLOOMOOLOO$ in the boxes, one to a box. To place the $L$'s we must choose 3 boxes from the 13. This can be done in $\binom{13}{3}$ ways. Now there are 10 boxes remaining and we choose one for the $M$. This can be done in $\binom{10}{1}$ ways. Out of the 9 remaining boxes we must choose 8 for the $O$'s and this can be done in $\binom{9}{8}$ ways. Then the $W$ is placed in the last box.

Thus the total number of arrangements is

$$\binom{13}{3}\binom{10}{1}\binom{9}{8}\binom{1}{1} = \frac{13!}{3!\,10!}\frac{10!}{1!\,9!}\frac{9!}{8!\,1!}$$

$$= \frac{13!}{3!\,1!\,8!\,1!}.$$

**Another way to think about this calculation**

Suppose that we first distinguish the letters by placing subscripts on them. Then we have 13 different symbols:

$$W, O_1, O_2, L_1, L_2, O_3, O_4, M, O_5, O_6, L_3, O_7 \text{ and } O_8.$$

Now there are 13! arrangements of these symbols. But many different permutations of the symbols correspond to the same arrangement of the underlying letters. In fact, there are 3! ways to permute the $L$'s and 8! ways to permute the $O$'s. Thus for every arrangement of the letters there are $3! \times 8!$ permutations of the subscripted letters which correspond to it. This means that the number of arrangements of the letters without subscripts is

$$\frac{13!}{3!\,8!},$$

as we saw before. ☐

**Arrangements involving indistinguishable objects**

The number of (ordered) arrangements of $n$ objects, in which there are $k_1$ objects of type 1, $k_2$ objects of type 2, ..., and $k_m$ objects of type $m$ and where $k_1 + k_2 + \cdots + k_m = n$, is

$$\frac{n!}{k_1!\,k_2!\,\ldots\,k_m!}.$$

This number is called a *multinomial coefficient*; it is written

$$\binom{n}{k_1, k_2, \ldots, k_m},$$

and pronounced "$n$ choose $k_1, k_2, \ldots, k_m$".

To see why the formula is true, imagine putting the $n$ objects in $n$ boxes. There are $\binom{n}{k_1}$ ways to choose the boxes in which to put the objects of type 1. This leaves $n - k_1$ boxes for the remaining objects and so there are $\binom{n-k_1}{k_2}$ ways to choose the boxes in which to put the objects of type 2. This leaves $n - k_1 - k_2$ boxes and so there are $\binom{n-k_1-k_2}{k_3}$ ways to choose the boxes for the third type of object. Continuing in this way, we see that the total number of arrangements is

$$\binom{n}{k_1}\binom{n-k_1}{k_2}\binom{n-k_1-k_2}{k_3}\cdots\binom{n-k_1-\cdots-k_{m-1}}{k_m}.$$

Writing out the binomial coefficients using formula (6.1) and cancelling, this expression reduces to

$$\frac{n!}{k_1!\,k_2!\,\ldots\,k_m!},$$

as required.

EXAMPLE 8.2 The total number of 10 digit numbers which can be formed from two 1's, two 2's, two 3's, two 4's and two 5's, is

$$\binom{10}{2,2,2,2,2} = \frac{10!}{2!\,2!\,2!\,2!\,2!} = \frac{10!}{(2!)^5}. \quad \square$$

EXAMPLE 8.3 The number of ways to divide 20 people into 4 committees each consisting of 5 people is

$$\frac{20!}{5!\,5!\,5!\,5!}. \quad \square$$

The *binomial* coefficient $\binom{n}{k}$ is a special case of a multinomial coefficient:

$$\binom{n}{k} = \binom{n}{k, n-k}.$$

EXAMPLE 8.4 Suppose a coin is tossed 24 times. Then the total number of possible outcomes (i.e., strings of heads and tails) is $2^{24}$ and the number of possible outcomes with 14 heads and 10 tails is

$$\frac{24!}{14!\,10!} = \binom{24}{14} = \binom{24}{10}. \quad \square$$

This sort of calculation can be combined with the Multiplication Principle:

EXAMPLE 8.5 Suppose a coin is tossed 24 times. How many possible outcomes are there with exactly 6 heads in the first 12 tosses?

The number of ways of having exactly 6 heads in the first 12 tosses is $\binom{12}{6}$. After the 12 tosses, there are two possibilities for the 13th toss, two for the 14th toss, ..., and two for the 24th toss. Hence the total number of possible outcomes is $\binom{12}{6} \times 2^{12} = 924 \times 2^{12}$. $\square$

## Multinomial theorem

A *multinomial* is an expression of the form

$$x_1 + x_2 + \cdots + x_m.$$

Suppose that we want to find the coefficient of $x_1^{k_1} x_2^{k_2} \ldots x_m^{k_m}$ in the expansion of

$$(x_1 + x_2 + \cdots + x_m)^n.$$

We write this expansion as
$$(x_1 + x_2 + \cdots + x_m)^n = (x_1 + x_2 + \cdots + x_m)$$
$$\times (x_1 + x_2 + \cdots + x_m)$$
$$\cdots$$
$$\times (x_1 + x_2 + \cdots + x_m)$$

The terms of the expansion are obtained by choosing one summand from each bracketed expression and multiplying them together. And so the coefficient of $x_1^{k_1} x_2^{k_2} \ldots x_m^{k_m}$ is simply the number of ways of choosing $k_1$ $x_1$'s, $k_2$ $x_2$'s, and so on, from the $n$ bracketed terms, one from each factor. This is the number of arrangements of $n$ symbols, $k_1$ of which are equal to $x_1$, $k_2$ of which are equal to $x_2$, etc. In other words, the coefficient of $x_1^{k_1} x_2^{k_2} \ldots x_m^{k_m}$ is
$$\binom{n}{k_1, k_2, \ldots, k_m}.$$

Thus we have the Multinomial Theorem.
$$(x_1 + x_2 + \cdots + x_m)^n = \sum_{k_1+k_2+\cdots+k_m=n} \binom{n}{k_1, k_2, \ldots, k_m} x_1^{k_1} x_2^{k_2} \ldots x_m^{k_m}.$$

EXAMPLE 8.6 The coefficient of $x_1^3 x_2^2 x_3^2 x_4^5$ in the expansion of $(x_1 + x_2 + x_3 + x_4)^{12}$ is
$$\binom{12}{3, 2, 2, 5} = \frac{12!}{3!\, 2!\, 2!\, 5!}. \quad \square$$

# Summary

After reading this chapter you should be able to calculate

- the number of permutations of objects, allowing repetitions;
- the coefficients in the expansion of $(x_1 + x_2 + \cdots + x_m)^n$.

# Problem Set 8

1. How many distinguishable arrangements are there of the letters in the words
   (*i*)   hodmandod,
   (*ii*)  imperseverant,
   (*iii*) myristicivorous,
   (*iv*)  indistinguishable,
   (*v*)   sociological,
   (*vi*)  Mississippi?

2. Suppose 20 people are divided into 6 different committees (labelled $C_1$ to $C_6$). Suppose that committee $C_1$ is to have 3 people, committee $C_2$ is to have 4 people, committee $C_3$ is to have 4 people, committee $C_4$ is to have 2 people, committee $C_5$ is to have 3 people and committee $C_6$ is to have 4 people. How many arrangements are there?

3. How many ways can the set $A = \{1, 2, 3, 4\}$ be written as the union of two disjoint subsets each of size 2? Write down all possibilities. For example $A = \{1, 2\} \cup \{3, 4\}$ is one of them.

4. In how many ways can 15 distinct balls be placed in 4 boxes so that the first box contains 5 balls, the second box contains 3 balls, the third box contains 4 balls and the fourth box contains 3 balls?

5. Suppose a single die is rolled 30 times, and the results are recorded in order. Suppose that the number 1 appeared 4 times, the number 2 appeared 2 times, the number 3 appeared 8 times, the number 4 appeared 5 times, the number 5 appears 4 times and the number 6 appeared 7 times. How many possibilities are there for the string of outcomes?

6. What is the coefficient of
   (i) $x_1^2 x_2 x_3$ in the expansion of $(x_1 + x_2 + x_3)^4$, and
   (ii) $x_1^2 x_2^3 x_3^2$ in the expansion of $(x_1 + x_2 + x_3)^7$?

7. Prove that for positive integers $m$ and $n$
$$m^n = \sum_{k_1+k_2+\cdots+k_m=n} \binom{n}{k_1, k_2, \ldots, k_m}.$$

*8. Show that the number of integer solutions to the equation
$$x_1 + 2x_2 + 3x_3 = 10,$$
where $0 \le x_1 \le 3$, $0 \le x_2 \le 4$ and $0 \le x_3 \le 5$ is the coefficient of $x^{10}$ in
$$(1+x+x^2+x^3)(1+x^2+x^4+x^6+x^8)(1+x^3+x^6+x^9+x^{12}+x^{15})$$
and thus find this number.

*9. (i) Given a set $A$ of size $n$. How many sequences of mutually disjoint subsets $A_1, A_2, \ldots, A_m$ are there whose union is $A$ and such that $|A_j| = k_j$ for $j = 1, 2, \ldots, m$?
   (ii) In how many ways can you write $\{1, 2, \ldots, 2n\}$ as a union of $n$ mutually disjoint subsets of equal size?

# 9

# Boolean Expressions

*George Boole (1815–1864)*

IN 1854 George Boole published his book: *An Investigation into the Laws of Thought, on Which are Founded the Mathematical Theories of Logic and Probability.* This was the beginning of the subject *Boolean algebra*: the algebra of set theory and logic which now has applications to switching circuits and digital electronics.

The connection with switching circuits is described in this chapter. Digital logic is the subject of Chapter 12.

*Turn back to the definitions in Chapter 2 to see how the words "and", "or" and "not" are used.*

The familiar algebra of arithmetic describes the rules obeyed by addition, multiplication and negation of numbers. In Boolean algebra we study the rules obeyed by "and", "or" and "not". It turns out that these are the same rules which apply to "intersection", "union" and "complement" in set theory. This should come as no surprise, because in the previous chapters you have already seen close connections between these ideas.

At the heart of this connection between the set theory studied in Chapter 2 and the mathematical approach to logic that you will find in Chapter 11 is the study of functions which take on only two values. In logic these two values are the symbols **TRUE** and **FALSE** but in other applications different symbols are used such as **ON** and **OFF**, or 1 and 0; all that matters is that there should be just two symbols.

*In set theory, the two symbols are 1 and 0.*

EXAMPLE 9.1 The subset $B$ of a set $A$ corresponds to the function $f : A \to \{0, 1\}$, where for all $x \in A$, $f(x) = 1$ if $x \in B$ and $f(x) = 0$ if $x \notin B$.

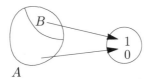

Figure 9.1
The function corresponding to a subset.

The sentence "$x$ belongs to $B$" is either true or false depending on the location of $x$. ❏

## Boolean functions

One way to obtain a two-valued function is to ask a yes/no question. Using "and", "or" and "not" we can combine simple yes/no questions into more complex ones. This leads naturally to the general study of Boolean functions. But rather than using **yes** and **no** as the two values we begin with 1 and 0. That is, we consider functions $f(x_1, \ldots, x_n)$ which take on only the values 1 and 0 and for which the variables take on only the values 1 and 0.

Let $S = \{1, 0\}$ and recall that

$$S^2 = S \times S = \{(1,1), (1,0), (0,1), (0,0)\},$$
$$S^3 = S \times S \times S$$
$$= \{(1,1,1), (1,1,0), (1,0,1), (1,0,0), (0,1,1),$$
$$(0,1,0), (0,0,1), (0,0,0)\},$$

and, in general

$$S^n = \underbrace{S \times S \cdots \times S}_{n}.$$

Then a function $f : S^n \to S$ is called a *Boolean function* of the $n$ variables $x_1, x_2, \ldots, x_n$.

The set $S^n$ has $2^n$ elements and therefore there are $2^{2^n}$ possible Boolean functions of $n$ variables.

## Decision tables

In order to illustrate these ideas we begin with a very simple situation.

**Problem 9.2** *At a particular time we ask the following three questions.*

(1)  *Is it cold?*
(2)  *Am I going out?*
(3)  *Is it raining?*

Let $x$, $y$ and $z$ represent the answers to these questions and let $f(x, y, z)$ represent the decision whether or not I wear a coat. Describe the function $f$.

**Solution.** We shall represent $f$ by a table. In the table, the variables $x$, $y$ and $z$ take on the values **y** or **n** according to whether the answer to the corresponding question is **yes** or **no**.

The actual values for $f(x, y, z)$ depend on your personal preferences. Here is our choice.

## Figure 9.2
A decision table

| $x$ | $y$ | $z$ | $f(x,y,z)$ |
|---|---|---|---|
| y | y | y | y |
| y | y | n | y |
| y | n | y | y |
| y | n | n | y |
| n | y | y | y |
| n | y | n | n |
| n | n | y | n |
| n | n | n | n |

In this table we have listed all possible combinations of $x$, $y$ and $z$. The values in the last column completely describe the function. Moreover, if we replace y by 1 and n by 0 we get a Boolean function.

*A Boolean function is completely described by its decision table.*

## Figure 9.3
A Boolean function

| $x$ | $y$ | $z$ | $f(x,y,z)$ |
|---|---|---|---|
| 1 | 1 | 1 | 1 |
| 1 | 1 | 0 | 1 |
| 1 | 0 | 1 | 1 |
| 1 | 0 | 0 | 1 |
| 0 | 1 | 1 | 1 |
| 0 | 1 | 0 | 0 |
| 0 | 0 | 1 | 0 |
| 0 | 0 | 0 | 0 |

❏

Decision tables can be a great help in clarifying the coding of decisions in computer programs. In particular, they ensure that all possible combinations of the conditions are accounted for.

## Switching circuits

Simple electrical circuits constructed only from switches have the property that current either flows or does not flow according to whether the various switches are on or off. Viewed in this way they are just like Boolean functions which use the values **on** and **off**.

*We could use 1 to mean the switch is on and 0 to mean that it is off.*

**Problem 9.3** *Can we build a simple circuit with a light and three switches labelled $x$, $y$ and $z$ so that by turning on the appropriate switch whenever the corresponding question is answered yes, the light will come on whenever I should wear my coat?*

For this to be possible we must allow each switch to open or close the circuit in several places if necessary. We label the places where switch $x$ operates with (x) and we use (x′) to denote a location which is open whenever $x$ is closed and closed whenever $x$ is open.

The following is the required circuit:

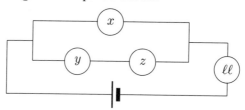

**Figure 9.4**
Check that the light comes on precisely when I should wear a coat!

We would like to know how to describe Boolean functions in terms of simple operations. This will help us translate these functions into switching circuits and *vice versa*.

First consider the simplest possible circuits. (From now on we leave out the lamp and the battery from our diagrams.)

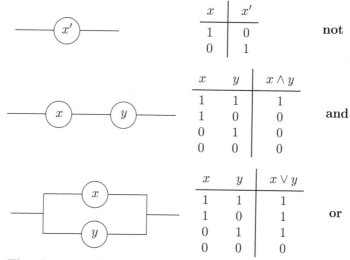

| $x$ | $x'$ |
|---|---|
| 1 | 0 |
| 0 | 1 |

**not**

| $x$ | $y$ | $x \wedge y$ |
|---|---|---|
| 1 | 1 | 1 |
| 1 | 0 | 0 |
| 0 | 1 | 0 |
| 0 | 0 | 0 |

**and**

| $x$ | $y$ | $x \vee y$ |
|---|---|---|
| 1 | 1 | 1 |
| 1 | 0 | 1 |
| 0 | 1 | 1 |
| 0 | 0 | 0 |

**or**

The above functions correspond to **not, and** and **or**. That is, in the diagram for $x'$, the current flows if and only if $x$ is *not* on. In the diagram for $x \wedge y$, the current flows if and only if $x$ *and* $y$ are on. In the diagram for $x \vee y$, the current flows if and only if $x$ *or* $y$ is on.

*We usually write $xy$ instead of $x \wedge y$.*

## Boolean expressions

An expression made up from variables such as $x$'s, $x'$'s using $\wedge$, $\vee$, 0 and 1 is called a *Boolean expression*.

The same symbols can be used to combine more complicated Boolean expressions. For example, the circuit for $fg$ is obtained by putting the circuits for $f$ and $g$ in *series* and the circuit for $f \vee g$ is obtained by putting the circuits for $f$ and $g$ in *parallel*.

## 62 BOOLEAN EXPRESSIONS

EXAMPLE 9.4 We can use Boolean expressions to describe Boolean functions. For example, the circuit (Figure 9.4) for the coat example corresponds to the Boolean expression

$$x \vee yz. \qquad \square$$

**Disjunctive normal form**

Here is how to obtain a Boolean expression from a table describing a Boolean function.

Suppose we have three variables $x$, $y$ and $z$. (The same method works for any number of variables.) For each line in the table for which the function has a 1, write down a product expression obtained by first taking $x$ if a 1 occurs in the $x$ column or by taking $x'$ if a 0 occurs in the $x$ column. Multiply this by $y$ if 1 occurs in the $y$ column or by $y'$ if 0 occurs in the $y$ column. Finally multiply by $z$ if 1 occurs in the $z$ column or by $z'$ if 0 occurs in the $z$ column. Then combine these expressions with $\vee$.

The Boolean expression thus obtained, is called the *disjunctive normal form* of the function.

EXAMPLE 9.5 The Boolean expression in disjunctive normal form for the Boolean function of Figure 9.3 is

$$xyz \vee xyz' \vee xy'z \vee xy'z' \vee x'yz.$$

**Equivalent Boolean expressions**

Many different Boolean expressions correspond to exactly the same Boolean function. For example the Boolean expressions

$$x \vee yz \quad \text{and} \quad xyz \vee xyz' \vee xy'z \vee xy'z' \vee x'yz$$

both correspond to the Boolean function given in Figure 9.3 for the "coat problem".

Boolean expressions corresponding to the same Boolean function are said to be *equivalent*. We shall write $f = g$ if the Boolean expressions $f$ and $g$ are equivalent.

EXAMPLE 9.6 The Boolean expressions $x \vee y$ and $(x \vee y) \vee x$ are equivalent: they have the same decision table.

| $x$ | $y$ | $x \vee y$ | $(x \vee y) \vee x$ |
|---|---|---|---|
| 1 | 1 | 1 | 1 |
| 1 | 0 | 1 | 1 |
| 0 | 1 | 1 | 1 |
| 0 | 0 | 0 | 0 |

# Laws of Boolean algebra

*E. V. Huntington (1874–1952)*

It turns out that two Boolean expressions are equivalent if and only if one can be obtained from the other by applying the following laws of Boolean algebra.

For all Boolean expressions $f$, $g$ and $h$, we have

*Actually, (ii) can be derived from the other laws.*

$(i)$ $\quad f \vee g = g \vee f, \quad\quad fg = gf,$
$(ii)$ $\quad (f \vee g) \vee h = f \vee (g \vee h), \quad (fg)h = f(gh),$
$(iii)$ $\quad f \vee (gh) = (f \vee g)(f \vee h), \quad f(g \vee h) = (fg) \vee (fh),$
$(iv)$ $\quad f \vee 0 = f, \quad\quad f \cdot 1 = f,$
$(v)$ $\quad f \vee f' = 1, \quad\quad ff' = 0.$

*These laws can also be checked directly using decision tables.*

An important consequence of these laws is the Boolean algebra version of de Morgan's laws:

$$(f \vee g)' = f'g',$$
$$(fg)' = f' \vee g'.$$

From these laws many others can be derived but, as you will see from the example, the derivations are not always obvious.

EXAMPLE 9.7 Here is how to prove that $ff = f$.

**Proof.** First use $(iv)$ and $(v)$ to write

$$f = f \cdot 1 = f(f \vee f').$$

Next use the distributive law $(iii)$ to obtain

$$f(f \vee f') = ff \vee ff'.$$

But from $(v)$ we have $ff' = 0$ and from $(iv)$ we have $ff \vee 0 = ff$. Therefore

$$f(f \vee f') = ff,$$

and it follows that $ff = f$. ❑

A similar proof shows that $f \vee f = f$.
We leave the derivation of other laws such as

$$f \vee 1 = 1, \quad \text{and } f \vee (fg) = f.$$

as a challenging exercise.

Another use for these laws is to reduce complicated Boolean expressions to simpler, more manageable forms.

EXAMPLE 9.8 Consider the Boolean expression
$$xyz \lor xyz' \lor xy'z \lor xy'z' \lor x'yz$$
representing the Boolean function of Figure 9.3. Using the rules of Boolean algebra, we have

$xyz \lor xyz' \lor xy'z \lor xy'z' \lor x'yz$
$= xyz \lor xyz' \lor xy'z \lor xy'z' \lor xyz \lor x'yz$
$= xy(z \lor z') \lor xy'(z \lor z') \lor (x \lor x')yz$
$= xy \lor xy' \lor yz$
$= x(y \lor y') \lor yz$
$= x \lor yz,$

so that
$$xy'z' \lor x'yz \lor xy'z \lor xyz' \lor xyz \quad \text{and} \quad x \lor yz$$
are equivalent. ❑

This method of reducing a Boolean expression to a simpler one can be quite tedious. In the next chapter, we shall discuss a systematic procedure which produces a simpler expression equivalent to the given one.

Yet another approach makes use of the fact that the laws of Boolean algebra have nice interpretations in terms of switching circuits. Therefore instead of using the rules of Boolean algebra to reduce an expression you can draw the corresponding circuit and simplify it directly.

EXAMPLE 9.9 The expression $xyz \lor xy'z$ is equivalent to $xz$. This can be seen directly from the circuit. In the following illustration, all three circuits have the same effect.

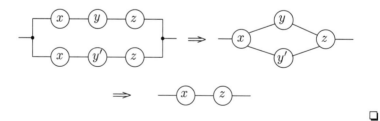

❑

## An unsolved problem

Given a Boolean expression in variables $x_1, x_2, \ldots, x_n$, how quickly can we determine whether the expression ever takes the value 1? At first sight the time appears to be proportional to $2^n$ (i.e., try every possible combination). We would like to find a method whose running time depends on a power of $n$ or else *prove* that this is not possible. This is an important unsolved problem.

# Summary

After reading this chapter you should be able

- write Boolean expressions in disjunctive normal form;
- design switching circuits for Boolean functions.

# Problem Set 9

1. Write a Boolean expression and the Boolean function for each of the following switching circuits.

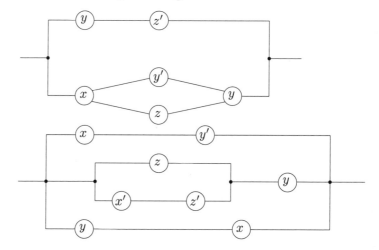

2. Write down a Boolean expression in disjunctive normal form which represents the following Boolean function and then find its corresponding switching circuit.

| $x$ | $y$ | $z$ | $f(x,y,z)$ |
|---|---|---|---|
| 1 | 1 | 1 | 0 |
| 1 | 1 | 0 | 1 |
| 1 | 0 | 1 | 1 |
| 1 | 0 | 0 | 1 |
| 0 | 1 | 1 | 0 |
| 0 | 1 | 0 | 1 |
| 0 | 0 | 1 | 0 |
| 0 | 0 | 0 | 1 |

3. Find a switching circuit corresponding to the following Boolean function

| $x$ | $y$ | $z$ | $f(x,y,z)$ |
|---|---|---|---|
| 1 | 1 | 1 | 0 |
| 1 | 1 | 0 | 1 |
| 1 | 0 | 1 | 1 |
| 1 | 0 | 0 | 0 |
| 0 | 1 | 1 | 1 |
| 0 | 1 | 0 | 0 |
| 0 | 0 | 1 | 0 |
| 0 | 0 | 0 | 0 |

4. Use the laws of Boolean algebra to prove that the Boolean expressions

$$xyz \vee xyz' \vee xy'z \vee x'yz \vee x'y'z' \vee x'y'z$$

and

$$z \vee xy \vee x'y'$$

are equivalent.

*5. Using the rules of Boolean algebra, prove the following identities for all Boolean expressions $f$ and $g$:

(i) $f \vee f = f$,  (ii) $ff = f$.
(iii) If $f \vee g = f$ for all $f$, then $g = 0$.
(iv) If $fg = f$ for all $f$, then $g = 1$.
(v) If $f \vee g = 1$ and $fg = 0$, then $g = f'$.
(vi) $f'' = f$,  (vii) $0' = 1$,  (viii) $1' = 0$.
(ix) $f \vee 1 = 1$,  (x) $f \cdot 0 = 0$.
(xi) $f \vee (fg) = f$,  (xii) $f(f \vee g) = f$.
(xiii) If $f \vee g = f \vee h$ and $f' \vee g = f' \vee h$, then $g = h$.
(xiv) $f(gh) = (fg)h$,  (xv) $f \vee (g \vee h) = (f \vee g) \vee h$.
(xvi) $(f \vee g)' = f'g'$,  (xvii) $(fg)' = f' \vee g'$.

*6. For subsets $A$ and $B$ of a set $U$, interpret $AB$ to mean $A \cap B$, $A \vee B$ to mean $A \cup B$, and $A'$ to mean $U \setminus A$. Show that the laws of Boolean algebra hold for the subsets of $U$. (You will have to give suitable interpretations to 0 and 1 — what are they?)

# 10

# Karnaugh Maps

IF we use the method described in the last chapter to produce a Boolean expression from a Boolean function, we obtain a list of *product terms* combined by $\vee$. This is called the *disjunctive normal form* of the function. Usually we would like a simpler expression which does the same job. That is, we want to minimize the number of switches in the corresponding circuit. One way we can attempt to do this is to use the laws of Boolean algebra to transform our expression into equivalent ones. But what we really need is a systematic procedure which produces simpler expressions. The method of *Karnaugh maps* is such a procedure.

## The Karnaugh map

*Notice that in moving from a column to an adjacent column, only one of the factors changes.*

The *Karnaugh map* of a Boolean function of $x$, $y$ and $z$ is a table with two rows and four columns. The rows are labelled $x$ and $x'$ and the columns are labelled $yz$, $yz'$, $y'z'$ and $y'z$. We put a 1 in a square of the table if the product of the row label and the column label occurs in the disjunctive normal form of the function. Elsewhere we put a 0. (Some books just leave these entries blank.)

EXAMPLE 10.1 In Example 9.5 the Boolean expression is

$$xyz \vee xyz' \vee xy'z \vee xy'z' \vee x'yz.$$

and the Karnaugh map (of the corresponding Boolean function) is

|    | $yz$ | $yz'$ | $y'z'$ | $y'z$ |
|----|------|-------|--------|-------|
| $x$  | 1    | 1     | 1      | 1     |
| $x'$ | 1    | 0     | 0      | 0     |

❑

If we have four variables, say $w$, $x$, $y$ and $z$, then the Karnaugh map has four rows, labelled $wx$, $wx'$, $w'x'$ and $w'x$, and has four columns, labelled $yz$, $yz'$, $y'z'$ and $y'z$.

EXAMPLE 10.2 For the Boolean expression

$wxyz \lor w'x'yz \lor w'x'yz' \lor w'x'y'z' \lor w'x'y'z \lor w'xyz' \lor w'xy'z' \lor w'xy'z$,

the Karnaugh map is

|      | $yz$ | $yz'$ | $y'z'$ | $y'z$ |
|------|------|-------|--------|-------|
| $wx$   | 1    | 0     | 0      | 0     |
| $wx'$  | 0    | 0     | 0      | 0     |
| $w'x'$ | 1    | 1     | 1      | 1     |
| $w'x$  | 0    | 1     | 1      | 1     |

□

## The Karnaugh map method

Here is the method which constructs a simple Boolean expression from the table. In the following description a *block* is a rectangular region of the Karnaugh map in which every entry is 1. We think of the top row of the table being adjacent to the bottom row and the right hand column being adjacent to the left hand column. That is, the table should really be drawn on a torus (i.e., on a doughnut). So a block can wrap around from left to right, or from top to bottom.

($i$)   Circle all blocks of eight 1's.
($ii$)  Circle all blocks of four 1's not wholly contained in blocks already marked.
($iii$) Circle all blocks of two 1's not wholly contained in blocks already marked.
($iv$)  Circle all blocks of one 1's not wholly contained in blocks already marked.
($v$)   Select as few of the outlined blocks as possible such that the selected blocks include every square which has a 1 in it.
($vi$)  For each block, write down the corresponding product term.
($vii$) Combine the product terms with $\lor$.

A variable, such as $x$ occurs in the product term for a block only if it is "constant on the block". That is, it occurs in the label of each row (or column) of the block and $x'$ does not occur. Conversely, $x'$ will be part of the product term if $x$ does not appear in the label of any row (or column) of the block. See Example 10.3 on the next page.

The essential point to note is that the product terms chosen in ($vi$) have a very simple form. On the other hand, there may be several different ways to choose the blocks in ($v$). That is, the Karnaugh map method does not always produce a unique simplest expression. Nevertheless, it usually produces an expression considerably simpler than the disjunctive normal form.

EXAMPLE 10.3 The block in the following Karnaugh map corresponds to the product term $x'y'$.

|      | $yz$ | $yz'$ | $y'z'$ | $y'z$ |
|------|------|-------|--------|-------|
| $wx$  | 0 | 0 | 0 | 0 |
| $wx'$ | 0 | 0 | 1 | 1 |
| $w'x'$| 0 | 0 | 1 | 1 |
| $w'x$ | 0 | 0 | 0 | 0 |

We can see that $x'y'$ is the product term, because only $x'$ and $y'$ are "constant on the block". Alternatively, we can calculate this directly using the laws of Boolean algebra:

$$wx'y'z \lor w'x'y'z \lor wx'y'z' \lor w'x'y'z'$$
$$= (w \lor w')x'y'z \lor (w \lor w')x'y'z'$$
$$= x'y'z \lor x'y'z'$$
$$= x'y'(z \lor z')$$
$$= x'y' \quad \square$$

**Problem 10.4** *Find a simple switching circuit corresponding to the Boolean function*

| $x$ | $y$ | $z$ | $f(x,y,z)$ |
|---|---|---|---|
| 1 | 1 | 1 | 1 |
| 1 | 1 | 0 | 1 |
| 1 | 0 | 1 | 1 |
| 1 | 0 | 0 | 0 |
| 0 | 1 | 1 | 0 |
| 0 | 1 | 0 | 1 |
| 0 | 0 | 1 | 0 |
| 0 | 0 | 0 | 1 |

**Solution.** Working directly from the table, a corresponding Boolean expression is

$$xyz \lor xyz' \lor xy'z \lor x'yz' \lor x'y'z'.$$

Thus a switching circuit (not the best one) corresponding to the given Boolean function is

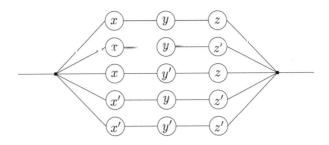

However we can obtain a simpler Boolean expression by using the Karnaugh map method as follows. We look for rectangular blocks of two, four or eight 1's. There are four such blocks (each with two 1's) but only three of them are necessary. The "split" block corresponds to $xz$, and the other two blocks that we have marked correspond to $xy$ and $x'z'$, respectively.

*We could have chosen $yz'$ instead of $xy$.*

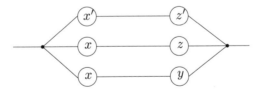

Therefore a simpler Boolean expression is

$$x'z' \vee xz \vee xy.$$

Hence a switching circuit (a better one!) corresponding to the given Boolean function is

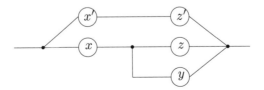

Even though the Karnaugh map method produces simple expressions, it does not always produce a circuit with the smallest number of switches. For example, the circuit just found can be simplified further:

*Even better than the Karnaugh map solution.*

□

EXAMPLE 10.5 Consider the Karnaugh map of the Boolean expression of the "coat" problem (Example 10.1). Using the Karnaugh map method, we see that a simpler Boolean expression which represents the same function is $x \vee yz$ (cf. Example 9.4).

|    | $yz$ | $yz'$ | $y'z'$ | $y'z$ |
|----|------|-------|--------|-------|
| $x$  | 1    | 1     | 1      | 1     |
| $x'$ | 1    | 0     | 0      | 0     |

□

EXAMPLE 10.6 The Boolean expression in disjunctive normal form for the Karnaugh map

|    | $yz$ | $yz'$ | $y'z'$ | $y'z$ |
|----|------|-------|--------|-------|
| $x$  | 1    | 1     | 0      | 1     |
| $x'$ | 1    | 0     | 1      | 1     |

is

$$xyz \vee xyz' \vee xy'z \vee x'yz \vee x'y'z' \vee x'y'z.$$

However we can obtain a simpler Boolean expression by using the Karnaugh map method. We look for rectangular blocks of four 1's, two 1's, etc. There is one "split" block of four (giving $z$), and two blocks of two (giving $xy$ and $x'y'$).

|    | $yz$ | $yz'$ | $y'z'$ | $y'z$ |
|----|------|-------|--------|-------|
| $x$  | 1    | 1     | 0      | 1     |
| $x'$ | 1    | 0     | 1      | 1     |

Thus a simpler Boolean expression is

$$z \vee xy \vee x'y'. \quad \square$$

EXAMPLE 10.7 Consider the Karnaugh map of Example 10.2.

|       | $yz$ | $yz'$ | $y'z'$ | $y'z$ |
|-------|------|-------|--------|-------|
| $wx$   | 1    | 0     | 0      | 0     |
| $wx'$  | 0    | 0     | 0      | 0     |
| $w'x'$ | 1    | 1     | 1      | 1     |
| $w'x$  | 0    | 1     | 1      | 1     |

We obtain a simple Boolean expression for this map by searching for rectangular blocks of eight 1's, four 1's, etc. There are three blocks of four and one block of one. The rectangular block of four gives rise to $w'x'$, and the other two square blocks of four give us the terms $w'z'$ and $w'y'$. The single "1" gives $wxyz$.

Hence a simpler Boolean expression is

$$wxyz \lor w'x' \lor w'y' \lor w'z'. \quad \square$$

*Sometimes there are several ways to simplify a Boolean expression.*

Remember that the Boolean expressions produced by the Karnaugh map method are not necessarily unique. The expression obtained depends on the selection of circled blocks and often there is more than one way to make this selection.

**Other methods**

The Karnaugh map method is only practical for three or four variables, although it can be used for five or six. Beyond six variables it is better to use another (equivalent) method: the Quine-McClusky algorithm. In the 1950's and 1960's Karnaugh maps and the Quine-McClusky method were used to minimize the number of electronic gates in digital logic circuits. These techniques are no longer used so extensively. Instead, circuits are built from data-selectors, programmable logic arrays or microprocessors. For example, a data-selector allows you to describe the decision table you need simply by setting certain inputs high or low.

# Summary

After reading this chapter you should be able to use Karnaugh maps to find simple forms for Boolean expressions and functions of three and four variables.

# Problem Set 10

1. Find a simple switching circuit corresponding to the following Boolean function

| $x$ | $y$ | $z$ | $f(x,y,z)$ |
|---|---|---|---|
| 1 | 1 | 1 | 0 |
| 1 | 1 | 0 | 1 |
| 1 | 0 | 1 | 1 |
| 1 | 0 | 0 | 1 |
| 0 | 1 | 1 | 1 |
| 0 | 1 | 0 | 1 |
| 0 | 0 | 1 | 0 |
| 0 | 0 | 0 | 0 |

2. Find a simple switching circuit corresponding to the following Boolean function

| $x$ | $y$ | $z$ | $g(x,y,z)$ |
|---|---|---|---|
| 1 | 1 | 1 | 1 |
| 1 | 1 | 0 | 1 |
| 1 | 0 | 1 | 0 |
| 1 | 0 | 0 | 0 |
| 0 | 1 | 1 | 1 |
| 0 | 1 | 0 | 1 |
| 0 | 0 | 1 | 1 |
| 0 | 0 | 0 | 0 |

3. For the following Karnaugh maps write down corresponding (simple) Boolean expressions.

(i)

|  | $yz$ | $yz'$ | $y'z'$ | $y'z$ |
|---|---|---|---|---|
| $x$ | 1 | 1 | 1 | 1 |
| $x'$ | 1 | 0 | 0 | 1 |

(ii)

|  | $yz$ | $yz'$ | $y'z'$ | $y'z$ |
|---|---|---|---|---|
| $wx$ | 1 | 1 | 1 | 0 |
| $wx'$ | 0 | 1 | 1 | 1 |
| $w'x'$ | 1 | 1 | 1 | 1 |
| $w'x$ | 1 | 1 | 1 | 0 |

(iii)

|  | $yz$ | $yz'$ | $y'z'$ | $y'z$ |
|---|---|---|---|---|
| $x$ | 1 | 0 | 1 | 1 |
| $x'$ | 1 | 1 | 0 | 1 |

(iv)

|  | $yz$ | $yz'$ | $y'z'$ | $y'z$ |
|---|---|---|---|---|
| $wx$ | 0 | 1 | 0 | 0 |
| $wx'$ | 0 | 1 | 1 | 1 |
| $w'x'$ | 1 | 1 | 1 | 0 |
| $w'x$ | 0 | 1 | 1 | 0 |

# 11

# Logic

AN important part of mathematics is establishing the truth or falsity of mathematical assertions such as "there are infinitely many prime numbers", or "every quadratic equation has two real roots". We shall see that we can think of these mathematical statements as Boolean expressions or Boolean functions defined using TRUE and FALSE instead of 1 and 0. This application of Boolean algebra is called the *logic of propositions* or the *propositional calculus*.

## Propositions

By *proposition* we mean a sentence that is either TRUE or FALSE.

It is often the case that the truth or falsity of a sentence, such as "It is raining.", depends on time. In order to regard these as propositions we implicitly assume that the time is the present.

EXAMPLE 11.1 The following are propositions:

(a) It is cold.
(b) Pigs have wings.
(c) Sydney is in Australia.
(d) It is always cold in Melbourne.
(e) $1 + 1 = 3$.
(f) $12^3 + 1 = 10^3 + 9^3$.
(g) The moon is a cube. ❏

EXAMPLE 11.2 The following are not propositions.

(*i*) $x = 3$.
(*ii*) How are you?
(*iii*) Have a cup of coffee. ❏

If $p$ and $q$ are propositions, we are interested in the way that the truth or falsity of propositions constructed from $p$ and $q$ depend on the truth or falsity of $p$ and $q$.

## Compound propositions

*Compound propositions* can be formed by combining $p$ and $q$ with **and, or, not, implies, if and only if** and other similar words.

EXAMPLE 11.3 The following are compound propositions:

(a) It is always cold in Sydney *and* $1 + 1 = 3$.
(b) It is cold *or* I am going out.
(c) It is *not* raining.
(d) The moon is a cube *if and only if* the sun rises in the west.
(e) *If* I am going out, *then* I shall wear a coat.
(f) The moon is a cube *implies* $1 + 1 = 3$.    ❏

Note that "if $p$, then $q$" has the same meaning as "$p$ implies $q$".

## Basic connectives

Instead of using words to connect propositions we shall use the following symbols:

| | | |
|---|---|---|
| $p \wedge q$ | means | $p$ **and** $q$ |
| $p \vee q$ | means | $p$ **or** $q$ |
| $\sim p$ | means | **not** $p$ |
| $p \Rightarrow q$ | means | $p$ **implies** $q$ (or **if** $p$, **then** $q$) |
| $p \Leftrightarrow q$ | means | $p$ **if and only if** $q$ |

Using these symbols (called *connectives*) we can build compound propositions. In Chapter 9, we used the term *Boolean expression* instead of *compound proposition*.

## Truth tables

We use *truth tables* to show the truth values of the five basic compound propositions. We use $T$ for TRUE and $F$ for FALSE.

Figure 11.1
Five basic truth tables

| $p$ | $q$ | $p \wedge q$ |
|---|---|---|
| $T$ | $T$ | $T$ |
| $T$ | $F$ | $F$ |
| $F$ | $T$ | $F$ |
| $F$ | $F$ | $F$ |

(a) $p \wedge q$

| $p$ | $q$ | $p \vee q$ |
|---|---|---|
| $T$ | $T$ | $T$ |
| $T$ | $F$ | $T$ |
| $F$ | $T$ | $T$ |
| $F$ | $F$ | $F$ |

(b) $p \vee q$

| $p$ | $\sim p$ |
|---|---|
| $T$ | $F$ |
| $F$ | $T$ |

(c) $\sim p$

| $p$ | $q$ | $p \Rightarrow q$ |
|---|---|---|
| T | T | T |
| T | F | F |
| F | T | T |
| F | F | T |

(d) $p \Rightarrow q$

| $p$ | $q$ | $p \Leftrightarrow q$ |
|---|---|---|
| T | T | T |
| T | F | F |
| F | T | F |
| F | F | T |

(e) $p \Leftrightarrow q$

Notice that $p \vee q$ is the *inclusive* or. That is, $p \vee q$ is TRUE when both $p$ and $q$ are TRUE.

A truth table is nothing other than the table of a Boolean function in which the values 1 and 0 have been replaced by $T$ and $F$. In fact, the tables for $\sim p$, $p \vee q$ and $p \wedge q$ have already occurred in Chapter 9 in connection with the simplest possible switching circuits.

*In Chapter 9 we used $p'$ instead of $\sim p$ and $pq$ instead of $p \wedge q$.*

We know from Chapter 9 that there are 16 different Boolean functions of two variables and therefore there are more truth tables than those listed above. We shall see some of the others in Chapter 12 when we study digital logic.

EXAMPLE 11.4 The truth table for $\sim(p \Rightarrow q)$ is

| $p$ | $q$ | $p \Rightarrow q$ | $\sim(p \Rightarrow q)$ |
|---|---|---|---|
| T | T | T | F |
| T | F | F | T |
| F | T | T | F |
| F | F | T | F |

❏

EXAMPLE 11.5 The truth table for $p \wedge \sim q$ is

| $p$ | $q$ | $\sim q$ | $p \wedge \sim q$ |
|---|---|---|---|
| T | T | F | F |
| T | F | T | T |
| F | T | F | F |
| F | F | T | F |

❏

EXAMPLE 11.6  The truth table for $(p \Rightarrow q) \wedge (q \Rightarrow r)$ is

| $p$ | $q$ | $r$ | $p \Rightarrow q$ | $q \Rightarrow r$ | $(p \Rightarrow q) \wedge (q \Rightarrow r)$ |
|---|---|---|---|---|---|
| $T$ | $T$ | $T$ | $T$ | $T$ | $T$ |
| $T$ | $T$ | $F$ | $T$ | $F$ | $F$ |
| $T$ | $F$ | $T$ | $F$ | $T$ | $F$ |
| $T$ | $F$ | $F$ | $F$ | $T$ | $F$ |
| $F$ | $T$ | $T$ | $T$ | $T$ | $T$ |
| $F$ | $T$ | $F$ | $T$ | $F$ | $F$ |
| $F$ | $F$ | $T$ | $T$ | $T$ | $T$ |
| $F$ | $F$ | $F$ | $T$ | $T$ | $T$ |

□

## Equivalent compound propositions

Two compound propositions are said to be *equivalent* if they have the same truth table.

EXAMPLE 11.7  The propositions of Examples 11.4 and 11.5 have the same truth table and therefore $\sim(p \Rightarrow q)$ and $p \wedge \sim q$ are equivalent.

EXAMPLE 11.8  By comparing the following truth table for the proposition $(p \Rightarrow q) \wedge (q \Rightarrow p)$ with the truth table for $p \Leftrightarrow q$ given in Figure 11.1 (e) we see that these propositions are equivalent.

| $p$ | $q$ | $p \Rightarrow q$ | $q \Rightarrow p$ | $(p \Rightarrow q) \wedge (q \Rightarrow p)$ |
|---|---|---|---|---|
| $T$ | $T$ | $T$ | $T$ | $T$ |
| $T$ | $F$ | $F$ | $T$ | $F$ |
| $F$ | $T$ | $T$ | $F$ | $F$ |
| $F$ | $F$ | $T$ | $T$ | $T$ |

□

## Tautology

A compound proposition that is always TRUE is called a *tautology* or a *theorem*.

EXAMPLE 11.9  The following propositions are tautologies:

(i)  $p \vee \sim p$.
(ii) $[(p \Rightarrow q) \wedge (q \Rightarrow r)] \Rightarrow (p \Rightarrow r)$.

It is easy to see that ($i$) is always true, but ($ii$) requires further explanation. We shall use a truth table. Let $s$ be the proposition $(p \Rightarrow q) \wedge (q \Rightarrow r)$. The truth table for $s$ appears in Example 11.6 and we use this to obtain the truth table for $s \Rightarrow (p \Rightarrow r)$:

| $p$ | $q$ | $r$ | $s$ | $p \Rightarrow r$ | $s \Rightarrow (p \Rightarrow r)$ |
|---|---|---|---|---|---|
| T | T | T | T | T | T |
| T | T | F | F | F | T |
| T | F | T | F | T | T |
| T | F | F | F | F | T |
| F | T | T | T | T | T |
| F | T | F | F | T | T |
| F | F | T | T | T | T |
| F | F | F | T | T | T |

Since every entry in the last column is $T$, the proposition in ($ii$) is a tautology. ❑

## Contradiction

A compound proposition that is always FALSE is called a *contradiction*.

EXAMPLE 11.10 The following propositions are contradictions:
($i$)  $p \wedge \sim p$.
($ii$) $(p \wedge q) \wedge (\sim p)$.

Again ($i$) is obvious and ($ii$) follows from the truth table:

| $p$ | $q$ | $p \wedge q$ | $\sim p$ | $(p \wedge q) \wedge (\sim p)$ |
|---|---|---|---|---|
| T | T | T | F | F |
| T | F | F | F | F |
| F | T | F | T | F |
| F | F | F | T | F |

❑

## Quantifiers

At this point it is good idea to introduce two new abbreviations. We read the symbol $\forall$ as "for all", and the symbol $\exists$ as "there exists" or as "for some".

The symbols $\forall$ and $\exists$ are called *quantifiers* and we can use them to turn statements involving a variable $x$ into propositions.

EXAMPLE 11.11 The statement "$x$ is an integer and $x$ is odd" is not a proposition because we don't know what $x$ is. But, the statement "$(\forall x \in \mathbb{R})$ $x$ is an integer and $x$ is odd" is a proposition (and FALSE). Similarly, the statement "$(\exists x \in \mathbb{R})$ $x$ is an integer and $x$ is odd" is a proposition (and TRUE). ❑

The study of propositions built up from simple statements involving quantifiers, variables and the connectives of the propositional calculus is called *predicate calculus*. It is at the heart of mathematics. At the very least it provides a convenient and precise language for the expression of mathematical ideas. For example, the definition of $B \subseteq A$, where $A$ and $B$ are sets, can be written as

$$(\forall x) \ (x \in B) \Rightarrow (x \in A).$$

### Universal Sets

We use the notation $p(x)$ to represent a statement that involves the variable $x$. In order to work with statements of this type we need to know something about the possible values of $x$. We take the point of view that in any particular problem there is some set $U$, called the *universal set*, and we consider only those statements $p(x)$ for which $x \in U$. That is, we interpret $\forall x$ to mean $\forall x \in U$ and $\exists x$ to mean $\exists x \in U$.

EXAMPLE 11.12 Let $U$ be the set of integers and $p(x)$ the statement "$x$ is odd". The statements

$$(\forall x)p(x) \quad \text{and} \quad (\exists x)p(x)$$

are propositions: the first one is FALSE and the second one is TRUE. ❑

The truth or falsity of a statement involving quantifiers depends on the choice of universal set.

EXAMPLE 11.13 Let $p(x)$ be the statement $x^2 + 1 = 0$. If the universal set $U$ is taken to be the set of integers, then the proposition $(\exists x)p(x)$ is false. On the other hand, if $U$ is taken to be the set of complex numbers, then $(\exists x)p(x)$ is true. ❑

### More about implication

The proposition $p \Rightarrow q$ is read "$p$ **implies** $q$" or "**if** $p$, **then** $q$". It is important to understand when this proposition is true and when it is false. Remember that propositions must be either TRUE or FALSE. Take the example:

"If it is raining, then I shall wear my coat."

Suppose I make this statement. Under what circumstances have I lied to you? There is only one case, namely, when it is raining but I am not wearing my coat. Looking at the truth table 11.1(d) for $\Rightarrow$, we see that this is what we would expect.

The truth table for $\Rightarrow$ describes how implication is used in mathematics. One of the most common causes of error in undergraduate mathematics is the misuse of $\Rightarrow$. It is important to realize that $p$ need not have anything to do with $q$.

EXAMPLE 11.14 Consider the following four propositions.

(a) If $1 + 1 = 2$, then $\dfrac{d}{dx} \sin x = \cos x$.

(b) If pigs have wings, then I shall pass the examination.

(c) If pigs have wings, then I shall fail the examination.

(d) If the moon is a cube, then the sun rises in the west.

We note that in each of the given compound propositions $p \Rightarrow q$, $p$ has nothing to do with $q$. In the first example, since both $p$ and $q$ are TRUE, the proposition is TRUE. For the remaining three, $p$ is always FALSE and therefore each one of these propositions is TRUE. ❏

Of course, in the examples that arise in practice the propositions $p$ and $q$ usually refer to related ideas.

EXAMPLE 11.15 The proposition

$$(\forall x \in \mathbb{Z})(x = 5 \Rightarrow x^2 = 25)$$

is TRUE. This is because when $x = 5$ is FALSE it doesn't matter whether or not $x^2 = 25$ is TRUE or FALSE; the compound proposition is TRUE in both cases. On the other hand, if $x = 5$ is TRUE, then it so happens that $x^2 = 25$ is also TRUE, and so the compound proposition is again TRUE.

On the other hand, the proposition

$$(\forall x \in \mathbb{Z})(x = 5 \text{ implies } x^2 = 36)$$

is FALSE. This is because if $x = 5$ is TRUE then $x^2$ must be equal to 25 which shows that $x^2 = 36$ is FALSE, and so the compound proposition is FALSE. ❏

## More about negation

*Some books use ¬p instead of ∼p.*

As indicated above, we use $\sim p$ to mean the negation of $p$. For example, if $p$ is the sentence "It is cold", then $\sim p$ is the sentence "It is not cold". (When dealing with Boolean expressions we usually write $p'$ instead of $\sim p$.)

Negating compound statements requires some care. Using truth tables we can easily verify the following equivalences, where we use "$\equiv$" to indicate that the propositions have the same truth table. The first two are known as *de Morgan's rules*.

(i) $\sim(p \vee q) \equiv (\sim p) \wedge (\sim q)$,
(ii) $\sim(p \wedge q) \equiv (\sim p) \vee (\sim q)$,
(iii) $\sim(p \Rightarrow q) \equiv p \wedge \sim q$,

Further care is needed with sentences involving $\forall$ and $\exists$. The rules are

(iv) $\sim(\forall x)p(x) \equiv (\exists x)(\sim p(x))$,
(v) $\sim(\exists x)p(x) \equiv (\forall x)(\sim p(x))$.

EXAMPLE 11.16 The negation of

"All diligent students are successful",

is

"Some diligent student is not successful". ❏

## Logical argument

If we know that $p \Rightarrow q$ is TRUE and if we also know that $p$ is TRUE, then we can conclude that $q$ is TRUE. This sort of reasoning occurs quite often in mathematics. In this situation we call $p$ the *hypothesis* and $q$ the *conclusion*.

It is often the case that we must *prove* that a proposition of the form $p \Rightarrow q$ is TRUE.

By looking at the truth table for $\Rightarrow$ we see that there is only one way that $p \Rightarrow q$ can be FALSE. This is when $p$ is TRUE, but $q$ is FALSE.

So to show that $p \Rightarrow q$ is always TRUE we must show that whenever $p$ is TRUE, then $q$ must be TRUE as well.

EXAMPLE 11.17 In Exercise 4 of Problem Set 2, you were asked to prove that for any sets $A$ and $B$,

$$A \cap B = A \Rightarrow A \subseteq B.$$

This can be done by considering the case when $A \cap B = A$ is TRUE and then showing that $A \subseteq B$ is TRUE. Note that the fact that $A \subseteq B$ is TRUE is equivalent to another implication, namely:

$$(\forall x)((x \in A) \Rightarrow (x \in B)). \quad \square$$

## Contrapositive

Using truth tables we can check that the propositions

$$p \Rightarrow q \quad \text{and} \quad \sim q \Rightarrow \sim p$$

are equivalent.

The proposition $\sim q \Rightarrow \sim p$ is called the *contrapositive* of $p \Rightarrow q$. Instead of proving $p \Rightarrow q$ is TRUE, sometimes it is easier to prove the contrapositive. This is usually called "*arguing by contradiction*". That is, if we assume that $q$ is FALSE and then show that $p$ is FALSE, we can conclude that $p \Rightarrow q$ is TRUE.

EXAMPLE 11.18 Take the universal set to be the set $\mathbb{Z}$ of integers. To prove that

$$(\forall x)((x^2 \text{ is even}) \Rightarrow (x \text{ is even}))$$

it is easier to prove the contrapositive:

$$(\forall x)((x \text{ is odd}) \Rightarrow (x^2 \text{ is odd})). \quad \square$$

## Converse

Note carefully that $p \Rightarrow q$ is *not* the same as $q \Rightarrow p$. The proposition $q \Rightarrow p$ is the *converse* of $p \Rightarrow q$ and the *converse* of $(\forall x)(p(x) \Rightarrow q(x))$ is $(\forall x)(q(x) \Rightarrow p(x))$.

EXAMPLE 11.19 Taking the universal set $U$ to be the set of all integers, if $p(x)$ is "$x = 1$" and $q(x)$ is "$x^2 = 1$", then the statement

$$(\forall x)(p(x) \Rightarrow q(x))$$

is TRUE but its converse

$$(\forall x)(q(x) \Rightarrow p(x))$$

is FALSE. $\quad \square$

# Logical puzzles

The principles described in this section can often be used to solve logical puzzles. Lewis Carroll invented many of these and the following example is due to him.

EXAMPLE 11.20

Babies are illogical. Nobody who can manage a crocodile is despised. Illogical persons are despised. Therefore babies cannot manage crocodiles.

In this example the universal set $U$ is the set of all people. Suppose

$$B(x) \quad \text{means} \quad \text{"}x \text{ is a baby"},$$
$$M(x) \quad \text{means} \quad \text{"}x \text{ can manage crocodiles"},$$
$$D(x) \quad \text{means} \quad \text{"}x \text{ is despised"},$$
$$I(x) \quad \text{means} \quad \text{"}x \text{ is illogical"},$$

then the above argument can be written as

$$(\forall x)((B(x) \Rightarrow I(x)) \wedge (I(x) \Rightarrow D(x)) \wedge (M(x) \Rightarrow \sim D(x)))$$
$$\Rightarrow (\forall x)(B(x) \Rightarrow \sim M(x)).$$

*See Example 11.9 (ii).* The contrapositive of $(M(x) \Rightarrow \sim D(x))$ is $(D(x) \Rightarrow \sim M(x))$ and using the fact that $(p \Rightarrow q) \wedge (q \Rightarrow r)$ implies $p \Rightarrow r$, it is not too hard to see that the argument is correct. Alternatively, a truth table could be used. ❑

# Summary

After reading this chapter you should know about

*The truth table for*
$\Rightarrow$ *is very important.*

- connectives and quantifiers;
- truth tables;
- logical argument and the converse and contrapositive of an implication;

*Including ∀ and ∃.*

- the rules for negation of propositions.

# Problem Set 11

1. Draw truth tables for the following propositions.
   (i)   $\sim(p \vee q)$.
   (ii)  $\sim p \wedge \sim q$.
   (iii) $p \vee p$.
   (iv)  $(\sim(p \vee q)) \vee (\sim p \wedge q) \vee p$.
   (v)   $p \wedge q \wedge r$.
   (vi)  $p \wedge (p \Rightarrow q)$.

2. Decide whether the following propositions are true or false:
   (i) If $1 + 1 = 2$, then $2 + 2 = 5$.
   (ii) If $1 + 1 \neq 2$, then pigs might fly.
   (iii) If $1 + 1 = 2$, then $2 + 3 = 5$.

3. Show that the following pairs of propositions are equivalent:
   (i) $(p \vee q) \Rightarrow r$; $(\sim p \wedge \sim q) \vee r$.
   (ii) $p \Rightarrow \sim q$; $q \Rightarrow \sim p$.
   (iii) $p \vee (q \wedge r)$; $(p \vee q) \wedge (p \vee r)$.
   (iv) $p \Rightarrow (q \vee r)$; $(p \Rightarrow q) \vee (p \Rightarrow r)$.

4. Show that each of the following propositions is a tautology:
   (i) $\sim(p \vee q) \vee (\sim p \wedge q) \vee p$. (ii) $((p \vee q) \wedge \sim q) \Rightarrow p$.

5. Show that each of the following propositions is a contradiction:
   (i) $(p \wedge q) \wedge \sim(p \vee q)$.  (ii) $(p \vee q) \wedge (\sim p \wedge \sim q)$.

6. Write down the truth table for the proposition

   $$(p \wedge \sim q) \Rightarrow \sim(p \Rightarrow q)$$

   and determine whether or not it is a tautology or a contradiction.

7. Write the following propositions in symbolic form:
   (i) All hungry crocodiles are not amiable.
   (ii) Some crocodiles, if not hungry, are amiable.

8. Taking the universal set to be the set $\mathbb{R}$ of all real numbers, determine the truth or falsity of the following sentences.
   (i) $(\forall x) \, ((x \in \mathbb{Z}) \Rightarrow (x^2 - x - 1 > 0))$.
   (ii) $(\exists x) \, ((x \in \mathbb{Z}) \wedge (x^2 - x - 1 > 0))$.
   (iii) $(\forall x) \, ((x^2 = 1) \Rightarrow (x = 1))$.
   (iv) $(\exists x) \, ((x^2 = 1) \wedge (x = 1))$.

*9. (i) Write down all 16 truth tables of two variables $p$ and $q$.
   (ii) For each truth table found in (i) find a compound proposition formed from $p$, $q$, $\wedge$, $\vee$ and $\sim$ which has the same table.

# 12

# Digital Logic

So far we have seen how Boolean functions occur in connection with decision tables, switching circuits, Karnaugh maps and the logic of propositions. In this section we study circuits that can be used to compute truth tables. Equivalently, we want to compute the value of a Boolean expression, given the values of the variables.

## Logic gates

*The gates used in practice can have more than two inputs.*

The building blocks of digital circuits are called *logic gates*. In this book a *gate* is a simple circuit that in general has two inputs and a single output. The circuit corresponding to a Boolean expression in $n$ variables will have $n$ inputs and a single output. We go back to our earlier notation, using 1 instead of $T$ and 0 instead of $F$. The seven most common logic gates are shown in Figure 12.2. These logic gates are readily available digital electronic devices (called *chips* or *integrated circuits*).

## From expressions to circuits

*An* inverter *is another name for a* NOT *gate.*

EXAMPLE 12.1 Here is how to build a circuit to represent the Boolean expression $x \vee x'y$.

We have input lines labelled $x$ and $y$. We obtain $x'$ by passing the $x$-input through an *inverter* and then we obtain $x'y$ by using an AND gate. Finally, we combine the output from the AND gate with $x$ using an OR gate. The final output represents the expression $x \vee x'y$.

Figure 12.1

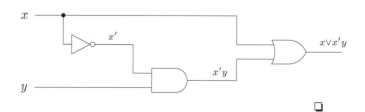

□

**86** DIGITAL LOGIC

Figure 12.2
Seven digital
logic gates

| Name | Logic Gate | Truth table |
|---|---|---|

**AND**

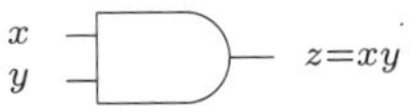

$x, y \rightarrow$ [AND gate] $\rightarrow z = xy$

| $x$ | $y$ | $z$ |
|---|---|---|
| 1 | 1 | 1 |
| 1 | 0 | 0 |
| 0 | 1 | 0 |
| 0 | 0 | 0 |

**OR**

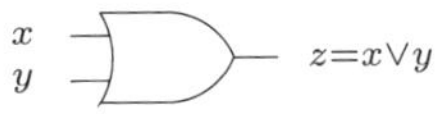

$x, y \rightarrow$ [OR gate] $\rightarrow z = x \vee y$

| $x$ | $y$ | $z$ |
|---|---|---|
| 1 | 1 | 1 |
| 1 | 0 | 1 |
| 0 | 1 | 1 |
| 0 | 0 | 0 |

A NOT *gate is an* inverter.

**NOT**

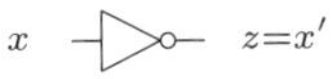

$x \rightarrow$ [NOT gate] $\rightarrow z = x'$

| $x$ | $z$ |
|---|---|
| 1 | 0 |
| 0 | 1 |

**NAND**

$x, y \rightarrow$ [NAND gate] $\rightarrow z = (xy)'$

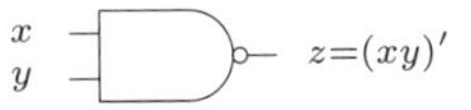

| $x$ | $y$ | $z$ |
|---|---|---|
| 1 | 1 | 0 |
| 1 | 0 | 1 |
| 0 | 1 | 1 |
| 0 | 0 | 1 |

**NOR**

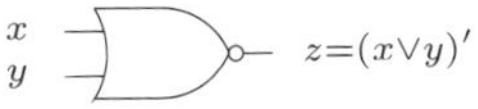

$x, y \rightarrow$ [NOR gate] $\rightarrow z = (x \vee y)'$

| $x$ | $y$ | $z$ |
|---|---|---|
| 1 | 1 | 0 |
| 1 | 0 | 0 |
| 0 | 1 | 0 |
| 0 | 0 | 1 |

*Exclusive OR* **XOR**

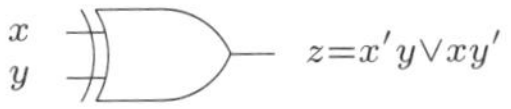

$x, y \rightarrow$ [XOR gate] $\rightarrow z = x'y \vee xy'$

| $x$ | $y$ | $z$ |
|---|---|---|
| 1 | 1 | 0 |
| 1 | 0 | 1 |
| 0 | 1 | 1 |
| 0 | 0 | 0 |

*Exclusive NOR* **XNOR**

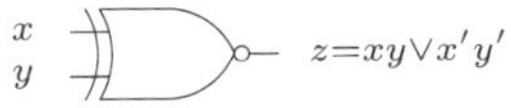

$x, y \rightarrow$ [XNOR gate] $\rightarrow z = xy \vee x'y'$

| $x$ | $y$ | $z$ |
|---|---|---|
| 1 | 1 | 1 |
| 1 | 0 | 0 |
| 0 | 1 | 0 |
| 0 | 0 | 1 |

DIGITAL LOGIC  **87**

EXAMPLE 12.2  Figure 12.3 is the digital logic circuit (using AND and OR gates and inverters) for the Boolean expression $x'y \vee xy'$.

Figure 12.3

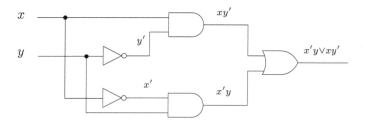

Note that the Boolean expression $x'y \vee xy'$ can be represented just by the XOR (exclusive OR) gate.  ❑

EXAMPLE 12.3  The following circuit represents the Boolean expression $(x \vee y)' \vee y'z$.

Figure 12.4

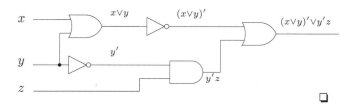

❑

## From circuits to expressions

EXAMPLE 12.4  Given a logic circuit, such as the one below, we can write down the Boolean expression which it represents by tracing the input lines through the gates.

Figure 12.5

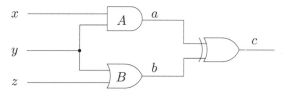

We use ⊕ as the symbol for the exclusive or.

The output of gate $A$ is $a = xy$ and the output of gate $B$ is $b = y \vee z$. The final output is the exclusive or $a \oplus b$ of $a$ and $b$. Thus the expression represented by the circuit is

$$xy \oplus (y \vee z).$$  ❑

EXAMPLE 12.5 The Boolean expression corresponding to the following digital logic circuit is $xy'z' \vee z$. As demonstrated in the previous example, we can see this simply be tracing the effect of the gates on the input lines (as indicated by the labels on the diagram).

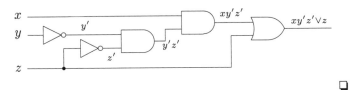

## Summary

After reading this chapter you should be able to

- construct digital circuits from Boolean expressions and *vice versa*.

## Problem Set 12

1. Draw the digital logic circuits, using AND and OR gates and inverters for the following Boolean expressions:

   (i) $xy \vee x'$     (ii) $xyz' \vee z$     (iii) $(x' \vee y)(x \vee y')$.

2. Write down the Boolean expression corresponding to the following logic circuit:

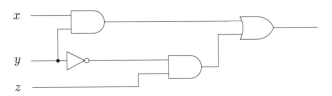

3. Construct a digital logic circuit which corresponds to the truth table for "implies".

*4. (i) Using only NOR gates, construct digital logic circuits which represent

   (a) NOT     (b) AND     (c) OR.

   This shows that every Boolean function can be constructed using only NOR gates.

   (ii) Repeat (i) using only NAND gates.

# 13

# Mathematical Induction

MATHEMATICAL induction is a general way of proving that a statement about the integer $n$ is true for all integers $n \geq k$, where $k$ is some fixed integer. The statement about $n$ is often deduced from observations of special cases. But no matter how many special cases we observe we cannot be sure that the statement is true *in general*, unless we *prove* it.

## The Principle of Mathematical Induction.

Let $S(n)$ be a statement about the integer $n$. Suppose that we know that for some integer $k$

(a)  $S(k)$ is true, and

*Generally k is a small integer such as 0 or 1.*

(b)  $(\forall n \geq k)(S(n) \Rightarrow S(n+1))$ is true.

Then we may conclude that $S(n)$ is true for all integers $n \geq k$.

This principle can be expressed as a statement about subsets of the set $\mathbb{N}$ of natural numbers:

Let $A \subseteq \mathbb{N}$. If $0 \in A$ and if $n \in A \Rightarrow n+1 \in A$, then $A = \mathbb{N}$.

To see the connection between this form of the principle of induction and the one above, take $A$ to be the set of all natural numbers $n$ for which $S(n+k)$ is true.

## Examples

The remainder of this chapter consists of examples of proofs using induction. In each case we have propositions $S(n)$ indexed by $n$ and we must show that $S(n) \Rightarrow S(n+1)$ is true. Look back to the truth table for "implies" (Figure 11.1 (d)) and you will see that the only way $S(n) \Rightarrow S(n+1)$ can be false is for $S(n)$ to be true but $S(n+1)$ to be false. We must show that this possibility never occurs. That is why in part (b) of each example we begin by saying "Suppose that $S(n)$ is true" and then deduce that $S(n+1)$ must be true also.

*Identity*

EXAMPLE 13.1 Use induction to prove that the following statement is true for all integers $n \geq 1$:
$$1 + 2 + 2^2 + 2^3 + \cdots + 2^n = 2^{n+1} - 1.$$
**Proof.** Let $S(n)$ be the given proposition. Then

(a) $S(1)$ is the proposition $1 + 2^1 = 2^{1+1} - 1$, which is clearly true.

(b) Suppose that $S(n)$ is true, i.e.
$$1 + 2 + 2^2 + 2^3 + \cdots + 2^n = 2^{n+1} - 1.$$
To prove that $S(n+1)$ is true, we proceed as follows:
$$1 + 2 + 2^2 + 2^3 + \cdots + 2^n + 2^{n+1}$$
$$= [1 + 2 + 2^2 + 2^3 + \cdots + 2^n] + 2^{n+1}$$
$$= 2^{n+1} - 1 + 2^{n+1} \quad \text{(hypothesis)}$$
$$= 2 \cdot 2^{n+1} - 1 = 2^{(n+1)+1} - 1.$$

Thus $S(n+1)$ is true and so we have shown that $S(n) \Rightarrow S(n+1)$ is true.

Hence $S(n)$ is true for all positive integers $n$. ∎

*Divisibility*

EXAMPLE 13.2 Use induction to show that $n^3 - 4n + 6$ is divisible by 3, for all positive integers $n$.

**Proof.** Let $S(n)$ be the proposition
$$n^3 - 4n + 6 \text{ is divisible by 3.}$$

(a) When $n = 1$,
$$n^3 - 4n + 6 = 1 - 4 + 6 = 3,$$
which is clearly divisible by 3. Hence $S(1)$ is true.

(b) Suppose that $S(n)$ is true, i.e.
$$n^3 - 4n + 6 = 3\ell, \quad \text{for some integer } \ell.$$
To prove that $S(n+1)$ is true, the following are the required steps:
$$(n+1)^3 - 4(n+1) + 6$$
$$= n^3 + 3n^2 + 3n + 1 - 4n - 4 + 6$$
$$= n^3 + 3n^2 - n + 3$$
$$= (n^3 - 4n + 6) + (3n^2 + 3n - 3)$$
$$= 3\ell + 3(n^2 + n - 1) \quad \text{(hypothesis)}.$$
and clearly the right-hand side is divisible by 3. Therefore $(n+1)^3 - 4(n+1) + 6$ is divisible by 3, i.e. $S(n+1)$ is true. Thus we have shown that $S(n) \Rightarrow S(n+1)$ is true.

Hence $S(n)$ is true for all positive integers $n$. ∎

## MATHEMATICAL INDUCTION

*Inequality* **EXAMPLE 13.3** Use induction to show that $2^n < n!$ for all integers $n > 3$.

**Proof.** Let $S(n)$ be the proposition "$2^n < n!$".

(a) An interesting point here is that the proposition $S(n)$ is false for $n = 1, 2, 3$. However, when $n = 4$, we see that $2^4 = 16$ and $4! = 24$ and so $2^4 < 4!$. Thus $S(4)$ is true.

(b) Suppose that $S(n)$ is true for some integer $n \geq 4$, i.e.,

$$2^n < n!.$$

To prove that $S(n+1)$ is true, we proceed as follows:

$$\begin{aligned} 2^{n+1} &= 2^n \cdot 2 \\ &< 2 \cdot n! \quad \text{(hypothesis)} \\ &< (n+1)n! \quad \text{(since } n \geq 4\text{)} \\ &= (n+1)!. \end{aligned}$$

Therefore $S(n+1)$ is true and so we have shown that $S(n) \Rightarrow S(n+1)$ is true.

Hence $S(n)$ is true for all $n > 3$. ☐

*Identity* **EXAMPLE 13.4** Use induction to prove that

$$1^3 + 2^3 + 3^3 + \cdots + n^3 = \frac{n^2(n+1)^2}{4},$$

for all positive integers $n$.

**Proof.** Let $S(n)$ be the given proposition. Then

(a) $S(1)$ is the proposition

$$1^3 = \frac{1^2(1+1)^2}{4},$$

which is clearly true.

(b) Suppose that $S(n)$ is true, i.e.

$$1^3 + 2^3 + 3^3 + \cdots + n^3 = \frac{n^2(n+1)^2}{4}.$$

To prove that $S(n+1)$ is true, we proceed as follows.

$$1^3 + 2^3 + 3^3 + \cdots + n^3 + (n+1)^3$$
$$= \left[1^3 + 2^3 + 3^3 + \cdots + n^3\right] + (n+1)^3$$
$$= \frac{n^2(n+1)^2}{4} + (n+1)^3 \quad \text{(hypothesis)}$$
$$= \frac{n^2(n+1)^2 + 4(n+1)^3}{4}$$
$$= \frac{(n+1)^2(n^2 + 4n + 4)}{4}$$
$$= \frac{(n+1)^2(n+2)^2}{4}$$
$$= \frac{(n+1)^2\bigl((n+1)+1\bigr)^2}{4}.$$

Thus $S(n+1)$ is true and so we have shown that $S(n) \Rightarrow S(n+1)$ is true. Hence $S(n)$ is true for all positive integers $n$. ∎

*Geometry* **EXAMPLE 13.5** What is the maximum number of regions in the plane defined by $n$ lines?

Let this number be $L_n$. By experiment we find that

$$L_0 = 1, \quad L_1 = 2, \quad L_2 = 4.$$

We are tempted to guess that $L_n = 2^n$. Using induction we can prove that $L_n \leq 2^n$, by observing that when we add a line it divides each region into at most two pieces. But $L_3 = 7$. In fact the $n$-th line splits $k$ old regions into two pieces if and only if it meets the previous lines in $k-1$ distinct points. So adding the $n$-th line to the picture can produce at most $n$ new regions. Thus $L_n \leq L_{n-1} + n$. But we guarantee that $n$ regions are split by making sure that the $n$-th line is not parallel to any of the previous lines and does not pass through any existing intersection. Then we can conclude that $L_n = L_{n-1} + n$. Now we see that the first few values of $L_n$ are 1, 2, 4, 7, 11, 16, .... Subtracting 1 from these numbers produces the series 0, 1, 3, 6, 10, 15, ... and if we are lucky we can recognize these numbers as binomial coefficients. Thus we conjecture that $L_n = 1 + \binom{n+1}{2} = \frac{1}{2}(n^2 + n + 2)$. Indeed, this can be proved by induction. The formula $L_n = L_{n-1} + n$ is an example of a recurrence relation. In Chapter 14 we shall use the method of generating functions to solve it. ∎

*Amusement*

**EXAMPLE 13.6** We shall attempt to prove by induction that "all horses are black".

Let $S(n)$ be the proposition that "any set of $n$ horses all have the same colour". Then $S(1)$ is certainly true. Now suppose that $S(n)$ is true. We will show that $S(n+1)$ is true.

Consider a paddock containing $n+1$ horses. Choose one of the horses and lead it away. This leaves $n$ horses and because $S(n)$ is true they all have the same colour. Now bring back the horse that was led away and choose another horse. Again because $S(n)$ is true the remaining $n$ horses all have the same colour. But this means that the horse that was chosen first has the same colour as the others. Thus $S(n+1)$ is true. It follows by induction that $S(n)$ is true for all $n \geq 1$.

*Where is the error in this argument?*

The other day I saw a black horse, therefore all horses are black! ❏

# Summary

After reading this chapter you should be able to use the Principle of Mathematical Induction to prove simple algebraic identities and inequalities.

# Problem Set 13

Use induction to prove the following propositions:

1. 6 divides $n(n^2 + 5)$ for all positive integers $n$.
2. $2^n \geq n + 10$, for all integers $n \geq 4$.
3. $1 + 3 + 5 + \cdots + (2n - 1) = n^2$, for all positive integers $n$.
4. The sum of the first $n$ positive even integers is $n^2 + n$.
5. $2 + 5 + 8 + \cdots + (3n-1) = \dfrac{n(3n+1)}{2}$, for all positive integers $n$.
6. 3 divides $n^3 + 5n$ for all $n \geq 1$.
7. $n! \geq 2^{n-1}$ for all positive integers $n$.
8. $11^n - 4^n$ is divisible by 7 for all positive integers $n$.
9. The sum of the cubes of any three consecutive positive integers is divisible by 9.
10. Prove that $5^n - 4n - 1$ is divisible by 16 for all positive integers $n$.
11. In Example 13.5, prove that $L_n = 1 + (1 + 2 + \cdots + n)$.

# 14

# Generating Functions

WHEN faced with a difficult counting problem, such as finding a formula for the $n$-th Catalan number $c_n$, it is often advisable to consider the numbers $c_0, c_1, c_2, \ldots$ simultaneously and to bundle them up into a single mathematical object. The object that we use is called a generating function and this way of tackling counting problems turns out to be the most powerful technique known.

Unfortunately "generating functions" are not functions at all and the use of the word "function" in this context is an historical accident. It would be much better to call them "formal power series", and some books do. We continue to call them "generating functions" because that is what you will find in most other books at this level.

In this chapter we describe the basic properties of generating functions and use them to solve some typical recurrence relations. In the next chapter we use the insights gained through these examples to give a more direct method of solving linear recurrence relations.

## The algebra of generating functions

Given a sequence

(14.1) $$a_0, a_1, a_2, a_3, \ldots,$$

its *generating function* is the power series

(14.2) $$G(z) = a_0 + a_1 z + a_2 z^2 + a_3 z^3 + \cdots.$$

In $\Sigma$-notation we write this as

$$G(z) = \sum_{k=0}^{\infty} a_k z^k.$$

EXAMPLE 14.3 The generating function of the Catalan numbers is

$$C(z) = \sum_{n=0}^{\infty} c_n z^n = 1 + z + 2z^2 + 5z^3 + 14z^4 + 42z^5 + \cdots \quad \square$$

**Addition and multiplication**  Generating functions can be added and multiplied just like polynomials. That is, if

$$G_1(z) = a_0 + a_1 z + a_2 z^2 + a_3 z^3 + \cdots$$

and

$$G_2(z) = b_0 + b_1 z + b_2 z^2 + b_3 z^3 + \cdots$$

then

$$G_1(z) + G_2(z) = (a_0 + b_0) + (a_1 + b_1)z + (a_2 + b_2)z^2 + (a_3 + b_3)z^3 + \cdots$$

and

$$G_1(z)G_2(z) = c_0 + c_1 z + c_2 z^2 + c_3 z^3 + \cdots,$$

where

$$c_k = a_0 b_k + a_1 b_{k-1} + a_2 b_{k-2} + \cdots + a_k b_0.$$

**Closed forms**  Generating functions are power series which usually involve infinitely many terms. We would like to find simpler expressions for these series, using only a finite number of mathematical operations. These simpler expressions are called *closed forms*.

If the terms of the sequence (14.1) are 0 from some point on, the generating function (14.2) is really an ordinary polynomial and it is already a closed form.

EXAMPLE 14.4  The generating function of the sequence

$$1,\ 6,\ 15,\ 20,\ 15,\ 6,\ 1,\ 0,\ 0,\ 0,\ \ldots$$

is

$$1 + 6z + 15z^2 + 20z^3 + 15z^4 + 6z^5 + z^6 = (1+z)^6. \quad \square$$

For many of the sequences (14.1), even those with infinitely many non-zero terms, we can still find closed forms.

EXAMPLE 14.5  The generating function of $1, 1, 1, 1, 1, 1, \ldots$ is

$$G(z) = 1 + z + z^2 + \cdots.$$

Multiplying by $1-z$ we find that $G(z)(1-z) = 1$ and this allows us to write

$$G(z) = \frac{1}{1-z}.$$

This is a closed form for $G(z)$.

Another way to see this is to observe that

$$G(z) - 1 = z + z^2 + z^3 + \cdots = z(1 + z + z^2 + \cdots) = zG(z).$$

Then $(1-z)G(z) = 1$ and again we find that $G(z) = (1-z)^{-1}$. □

## Negative powers of $1-z$

Later in this chapter it will be useful to know the coefficient of $z^m$ in $(1-z)^{-n}$. One way to find a formula for these coefficients is to raise the generating function of the previous example to the power $n$ and then to use some counting techniques from Chapter 6 to complete the calculation.

From Example 14.5 we have

$$(1-z)^{-1} = 1 + z + z^2 + z^3 + \cdots$$

and therefore, for $n \geq 1$,

$$(1-z)^{-n} = \underbrace{(1 + z + z^2 + z^3 + \cdots) \ldots (1 + z + z^2 + z^3 + \cdots)}_{n \text{ factors}}.$$

The coefficient of $z^m$ in this product is equal to the number of ways one can form $z^m$ by multiplying together $n$ terms, one from each factor. That is, it is the number of choices of $d_1, d_2, \ldots, d_n$ such that

$$z^{d_1} z^{d_2} \ldots z^{d_n} = z^m.$$

This is the number of solutions to the equation

$$d_1 + d_2 + \cdots + d_n = m,$$

where each $d_i$ is a non-negative integer. We know from Problem 6.12 of Chapter 6 that this number is

$$\binom{m+n-1}{m}.$$

Putting these observations together gives the formula

(14.6) $$(1-z)^{-n} = \sum_{m=0}^{\infty} \binom{m+n-1}{m} z^m.$$

EXAMPLE 14.7 When $n = 2$ we have

$$\binom{m+n-1}{m} = \binom{m+1}{m} = m+1,$$

and so

$$(1-z)^{-2} = 1 + 2z + 3z^2 + 4z^3 + \cdots. \quad \square$$

# Recurrence relations

A *recurrence relation* for a sequence
$$x_0, x_1, x_2, \ldots$$
is an expression for $x_n$ in terms of $x_0, x_1, \ldots, x_{n-1}$.

EXAMPLE 14.8 The Catalan numbers $c_n$ satisfy the recurrence relation

(14.9) $\qquad c_{n+1} = c_0 c_n + c_1 c_{n-1} + \cdots + c_n c_0.$

This will be proved in Chapter 19. If we know suitable starting values (in this case $c_0 = 1$) we can use the recurrence relation to calculate the sequence to as many terms as we please. For the Catalan numbers we find:
$$1, 1, 2, 5, 14, 42, \ldots \quad \square$$

We shall see that it is often the case that a recurrence relation can be translated into a formula involving the generating function of the sequence and from this we can obtain a closed form.

EXAMPLE 14.10 Let $L_n$ be the maximum number of regions in the plane produced by $n$ lines (Example 13.5). Then the $L_n$ satisfy the recurrence relation
$$L_n = L_{n-1} + n \quad \text{for } n \geq 1,$$
with $L_0 = 1$.

The generating function for the sequence $L_0, L_1, L_2, \ldots$ is
$$L(z) = L_0 + L_1 z + L_2 z^2 + L_3 z^3 + \cdots.$$
After multiplying this by $z$ we find that
$$zL(z) = \qquad L_0 z + L_1 z^2 + L_2 z^3 + \cdots.$$
Now we can subtract $zL(z)$ from $L(z)$ to obtain
$$L(z) - zL(z) = L_0 + (L_1 - L_0)z + (L_2 - L_1)z^2 + (L_3 - L_2)z^3 + \cdots.$$
But we know that $L_0 = 1$ and that $L_n - L_{n-1} = n$, therefore
$$(1-z)L(z) = 1 + z + 2z^2 + 3z^3 + \cdots$$
$$= 1 + z(1 + 2z + 3z^2 + \cdots).$$
Using Example 14.7 we see that
$$(1-z)L(z) = 1 + z(1-z)^{-2}.$$
Therefore a closed form for $L(z)$ is
$$L(z) = (1-z)^{-1} + z(1-z)^{-3}. \quad \square$$

Whenever we have a closed form, we can expand it to find the coefficients of the generating function.

EXAMPLE 14.11 Equation (14.6) allows us to expand the closed form of $L(z)$ that we found in the previous example. Thus

$$L(z) = \sum_{n=0}^{\infty} z^n + z \left( \sum_{m=0}^{\infty} \binom{m+2}{2} z^m \right)$$

$$= \sum_{n=0}^{\infty} z^n + \sum_{m=0}^{\infty} \binom{m+2}{2} z^{m+1}$$

$$= \sum_{n=0}^{\infty} z^n + \sum_{n=1}^{\infty} \binom{n+1}{2} z^n,$$

where we have used the substitution $n = m + 1$ in the second summation of the last line. By definition, the coefficient of $z^n$ is $L_n$ and therefore $L_n = 1 + \binom{n+1}{2}$. ❏

## Partial fraction expansions

It turns out that a large class of recurrence relations lead to generating functions that have a closed form which is a rational function; i.e., one polynomial divided by another.

EXAMPLE 14.12 The expression

$$\frac{1 + z + 3z^3}{2 + 5z - 4z^2}$$

is a rational function. ❏

The theory of partial fractions allows us to write any rational function as a sum of a polynomial and simple rational functions of the form

$$\frac{A}{(1 - \lambda z)^k}.$$

*λ may be a complex number.*

These simple rational functions can be expanded using the formula (14.6). After adding these expansions, the coefficient of $z^n$ gives the solution to the original recurrence relation.

EXAMPLE 14.13 Consider the recurrence relation:

$$x_n + 2x_{n-1} - 15x_{n-2} = 0,$$

for $n \geq 2$, where $x_0 = 0$ and $x_1 = 1$. Let

$$G(z) = x_0 + x_1 z + x_2 z^2 + x_3 z^3 + \cdots$$

be the generating function of the sequence $x_0, x_1, x_2, \ldots$. After multiplying by $z$ and $z^2$ we find that

$$zG(z) = \quad x_0 z + x_1 z^2 + x_2 z^3 + \cdots$$
$$z^2 G(z) = \quad x_0 z^2 + x_1 z^3 + \cdots$$

We now form the expression $G(z) + 2zG(z) - 15z^2 G(z)$. Because of the recurrence relation most of the terms cancel and so

$$G(z)(1 + 2z - 15z^2) = x_0 + (x_1 + 2x_0)z.$$

Using the initial values $x_0 = 0$ and $x_1 = 1$ and dividing by $(1 + 2z - 15z^2)$ leads to the closed form

$$G(z) = \frac{z}{1 + 2z - 15z^2}.$$

In order to find a formula for $x_n$ we use partial fractions to simplify the expression just found for $G(z)$.

We have $1 + 2z - 15z^2 = (1 - 3z)(1 + 5z)$ and we want to find numbers $A$ and $B$ such that

$$\frac{z}{1 + 2z - 15z^2} = \frac{A}{1 - 3z} + \frac{B}{1 + 5z}.$$

Solving this for $A$ and $B$ we find that $A = \frac{1}{8}$ and $B = -\frac{1}{8}$. Thus

$$G(z) = \frac{1}{8} \frac{1}{1 - 3z} - \frac{1}{8} \frac{1}{1 + 5z}.$$

Using the formula for $(1 - z)^{-1}$, replacing $z$ by $3z$ for the first term, and $z$ by $-5z$ for the second term, we obtain

$$G(z) = \frac{1}{8}(1 + 3z + 3^2 z^2 + \cdots) - \frac{1}{8}(1 - 5z + 5^2 z^2 - 5^3 z^3 + \cdots)$$

The coefficient of $z^n$ in $G(z)$ is $x_n$ and from the expression for $G(z)$ just found we have the solution

$$x_n = \frac{1}{8} \cdot 3^n - \frac{1}{8}(-5)^n$$

to our recurrence relation. □

# Summary

After reading this chapter you should know the meaning of

- generating function;
- closed form;
- recurrence relation.

## Problem Set 14

1. Find the generating functions for the following sequences and write them in as simple a form as possible.

   (i) The binomial coefficients $\binom{n}{0}, \binom{n}{1}, \ldots, \binom{n}{n}$.

   (ii) $1, a, a^2, a^3, \ldots$

2. Find a closed form for the generating function of the sequence $a_0, a_1, \ldots$, where $a_0 = 1$, $a_1 = 1$ and
$$a_n = 5a_{n-1} - 6a_{n-2}.$$

3. Let $a_0, a_1, a_2, \ldots$, be a sequence of integers satisfying the recurrence relation
$$a_n - 2a_{n-1} - a_{n-2} + 2a_{n-3} = 0,$$
for $n \geq 4$, where $a_0 = 0$, $a_1 = 0$, $a_2 = 2$ and $a_3 = 5$. Find a closed form for the generating function of this sequence.

4. Solve the following recurrence relations using generating functions:

   (i) $x_n = 4x_{n-1} - 3x_{n-2}$, where $x_0 = 1$ and $x_1 = 2$.

   (ii) $x_n = 3x_{n-1} - 3x_{n-2} + x_{n-3}$, where $x_0 = 0$, $x_1 = 1$ and $x_2 = 3$.

   (iii) $x_n = 10x_{n-1} - 25x_{n-2}$, for $n \geq 2$, where $x_0 = -1$ and $x_1 = 5$.

   (iv) $x_{n+3} - 6x_{n+2} + 11x_{n+1} - 6x_n = 0$, for $n \geq 0$, where $x_0 = 1$, $x_1 = 0$ and $x_2 = -1$.

5. Consider the recurrence relation
$$a_n = a_{n-1} + 4,$$
for $n > 0$ and where $a_0 = 1$. Find a closed form for its generating function and thereby find a formula for $a_n$.

*6. Consider the following recurrence relation
$$a_{n+2} = a_{n+1} + a_n + n$$
for $n \geq 0$ where $a_0 = a_1 = 1$. Write the corresponding generating function in its closed form.

# 15

# Linear Recurrence Relations

IN Chapter 14 we introduced recurrence relations and solved some typical examples using generating functions. In this chapter we continue to study recurrence relations but concentrate on a simple but important case which can be solved directly without recourse to generating functions. Nevertheless, we use generating functions to guide us to the correct form of the solution.

## Homogeneous linear recurrence relations

A *k-th order linear recurrence relation* for the sequence $x_0, x_1, \cdots$ has the form

$$x_n = a_1 x_{n-1} + a_2 x_{n-2} + \cdots + a_k x_{n-k} + f_n, \qquad \text{for } n \geq k,$$

where $a_1, a_2, \ldots, a_k$ are constants and $f_k, f_{k+1}, f_{k+2}, \ldots$ is some given sequence.

*Recurrence relations are called difference equations by some authors.*

We shall restrict our attention to the *homogeneous* linear recurrence relations: those for which all the $f_n$ are 0. These are the recurrence relations of the form

(15.1) $\qquad x_n = a_1 x_{n-1} + a_2 x_{n-2} + \cdots + a_k x_{n-k}, \qquad \text{for } n \geq k.$

or, equivalently,

(15.1*) $\qquad x_n - a_1 x_{n-1} - a_2 x_{n-2} - \cdots - a_k x_{n-k} = 0, \qquad \text{for } n \geq k.$

*Fibonacci numbers*  The sequence

$$1, 1, 2, 3, 5, 8, 13, 21, 34, 55, \ldots$$

in which each number is the sum of the two preceding numbers arises in a surprising number of places. This is the sequence of *Fibonacci numbers*. The next example describes one way of obtaining it.

**Problem 15.2** *Suppose that you can climb a staircase by taking either one or two steps at a time. Find a recurrence relation for the number of ways you can get to the n-th step.*

**Solution.** Let $F_n$ be the number of ways to get to the $n$-th step. Observe that you can get to the $n$-th step in two ways: either by taking a single stride from the $(n-1)$-st step or a single stride from the $(n-2)$-nd step. There are $F_{n-1}$ ways to get to the $(n-1)$-st step and $F_{n-2}$ ways to get to the $(n-2)$-nd step. Therefore we obtain the recurrence relation

$$F_n = F_{n-1} + F_{n-2}.$$

Because $F_0 = F_1 = 1$, this recurrence relation produces the sequence of Fibonacci numbers. ❑

Similarly, if $T_n$ is the number of ways to get to the $n$-th step when you can take up to three steps at a time, then the recurrence relation would be

$$T_n = T_{n-1} + T_{n-2} + T_{n-3}.$$

## First order linear recurrence relations

The simplest example of (15.1) is $x_n = \lambda x_{n-1}$. (In this case $k = 1$ and $a_1 = \lambda$.) You will find that the closed form for its generating function is $x_0(1 - \lambda z)^{-1}$ and you could use this to find $x_n$. On the other hand, we can find $x_n$ directly as follows.

Putting $n$ equal to 1, 2, 3, ... we find that

$$x_1 = \lambda x_0$$
$$x_2 = \lambda x_1 = \lambda^2 x_0$$
$$x_3 = \lambda x_2 = \lambda^3 x_0$$
$$\cdots$$

It is easy to see that, in general,

$$x_n = \lambda^n x_0.$$

Indeed we can prove this by induction. For $n = 1$ we have $x_1 = \lambda x_0$ by assumption. On the other hand, if $n \geq 1$ and $x_n = \lambda^n x_0$, then $x_{n+1} = \lambda x_n = \lambda \lambda^n x_0 = \lambda^{n+1} x_0$, completing the induction step.

# Second order linear recurrence relations

The previous example gives us a hint as to how to tackle the general case. Better understanding can be obtained by studying the second order recurrence relation

(15.3) $\qquad x_n = ax_{n-1} + bx_{n-2}, \qquad$ for $n \geq 2$.

If $x'_0, x'_1, x'_2, \ldots$ and $x''_0, x''_1, x''_2, \ldots$ are two sequences which satisfies the recurrence relation (15.3) and if $A$ and $B$ are constants, then $x_n = Ax'_n + Bx''_n$ also satisfies (15.3). This is because

$$\begin{aligned} x_n = Ax'_n + Bx''_n &= A(ax'_{n-1} + bx'_{n-2}) + B(ax''_{n-1} + bx''_{n-2}) \\ &= a(Ax'_{n-1} + Bx''_{n-1}) + b(Ax'_{n-2} + Bx''_{n-2}) \\ &= ax_{n-1} + bx_{n-2}. \end{aligned}$$

Thus the linear combination $(Ax'_n + Bx''_n)$ of the two solutions $(x'_n)$ and $(x''_n)$ of (15.3) is also a solution. This is a very important property and holds in general for all homogeneous linear recurrence relations.

**Problem 15.4** *What is the general solution to the recurrence relation (15.3)?*

**Solution.** Taking our cue from Example 14.13 we introduce the generating function $G(z)$ of $x_0, x_1, x_2, \ldots$ and observe that

$$\begin{aligned} G(z) &= x_0 + x_1 z + x_2 z^2 + x_3 z^3 + \cdots \\ zG(z) &= \phantom{x_0 +\,} x_0 z + x_1 z^2 + x_2 z^3 + \cdots \\ z^2 G(z) &= \phantom{x_0 + x_1 z +\,} x_0 z^2 + x_1 z^3 + \cdots. \end{aligned}$$

Therefore, using (15.3),

$$G(z) - azG(z) - bz^2 G(z) = x_0 + (x_1 - ax_0)z.$$

From this we obtain the closed form

$$G(z) = \frac{x_0 + (x_1 - ax_0)z}{1 - az - bz^2}.$$

The next step is to factorize $1 - az - bz^2$; i.e., write

$$1 - az - bz^2 = (1 - \lambda_1 z)(1 - \lambda_2 z)$$

for some numbers $\lambda_1$ and $\lambda_2$. We can now find a partial fraction expansion for $G(z)$. If $\lambda_1 \neq \lambda_2$, this will be of the form

$$G(z) = \frac{A}{1 - \lambda_1 z} + \frac{B}{1 - \lambda_2 z},$$

for some constants $A$ and $B$; whereas if $\lambda_1 = \lambda_2$, it will be of the form
$$G(z) = \frac{A}{1 - \lambda_1 z} + \frac{B}{(1 - \lambda_1 z)^2}.$$
In the first case, the coefficient of $z^n$ in $G(z)$ is
$$x_n = A\lambda_1^n + B\lambda_2^n$$
and in the second case (see Example 14.7), the coefficient of $z^n$ is
$$x_n = A\lambda_1^n + B(n+1)\lambda_1^n = (A+B)\lambda_1^n + Bn\lambda_1^n.$$
To find the values of $A$ and $B$ we need to know some *initial conditions* such as the values of $x_0$ and $x_1$. ☐

The calculations just completed show that the recurrence relation (15.3) always has a solution of the form $x_n = \lambda^n$ for some $\lambda$. Knowing this we can dispense with generating functions and find $\lambda$ directly. This is summarised in the following theorem and is best illustrated by example.

$\lambda_1$ and $\lambda_2$ are the roots of the characteristic equation $\lambda^2 - a\lambda - b = 0$.

**Theorem 15.5** *If $1 - az - bz^2 = (1 - \lambda_1 z)(1 - \lambda_2 z)$, then the general solution to the recurrence relation*
$$x_n = ax_{n-1} + bx_{n-2}, \qquad \text{for } n \geq 2.$$
*is either $x_n = A\lambda_1^n + B\lambda_2^n$ or $x_n = C\lambda_1^n + Dn\lambda_1^n$ according to whether $\lambda_1 \neq \lambda_2$ or $\lambda_1 = \lambda_2$, respectively.*

EXAMPLE 15.6 We shall solve the recurrence relation

Now we obtain the solution without using generating functions.

(15.7) $$x_n + 2x_{n-1} - 15x_{n-2} = 0,$$
for $n \geq 2$, where $x_0 = 0$ and $x_1 = 1$, which we solved in Example 14.13 using generating functions.

Substituting $x_n = \lambda^n$ into (15.7) gives
$$\lambda^n + 2\lambda^{n-1} - 15\lambda^{n-2} = 0.$$
When $n = 2$ this becomes
$$\lambda^2 + 2\lambda - 15 = 0,$$
and in factorized form this is $(\lambda - 3)(\lambda + 5) = 0$. Hence $\lambda$ is 3 or $-5$. Therefore, the general solution is
$$x_n = A\,3^n + B(-5)^n.$$
In order to find $A$ and $B$ we must use the initial conditions $x_0 = 0$ and $x_1 = 1$. Putting $n = 0$ and $n = 1$ leads to the equations
$$A + B = 0$$
$$3A - 5B = 1.$$
From these equations we have $A = \frac{1}{8}$ and $B = -\frac{1}{8}$. Hence the solution is
$$x_n = \frac{1}{8} \cdot 3^n - \frac{1}{8}(-5)^n. \quad \square$$

**Fibonacci numbers again**

At the beginning of this chapter we defined the Fibonacci numbers as the sequence 1, 1, 2, 3, 5, 8, ... satisfying the recurrence relation

$$F_n = F_{n-1} + F_{n-2},$$

with initial conditions $F_0 = F_1 = 1$.

We now look for a solution of the form $F_n = \lambda^n$. That is, we must have

$$\lambda^n = \lambda^{n-1} + \lambda^{n-2}.$$

In particular, for $n = 2$ we have $\lambda^2 = \lambda + 1$ and this quadratic equation has two solutions:

$$\lambda_1 = \tfrac{1}{2}(1 + \sqrt{5}) \quad \text{and} \quad \lambda_2 = \tfrac{1}{2}(1 - \sqrt{5}).$$

This leads to two solutions for $F_n$, namely

$$F_n = \left(\tfrac{1}{2}(1 + \sqrt{5})\right)^n \quad \text{and} \quad F_n = \left(\tfrac{1}{2}(1 - \sqrt{5})\right)^n.$$

Neither of the solutions just found satisfies $F_1 = 1$. However, the general solution

$$F_n = A\left(\frac{1+\sqrt{5}}{2}\right)^n + B\left(\frac{1-\sqrt{5}}{2}\right)^n$$

will satisfy the initial conditions provided we choose the correct values for $A$ and $B$. Putting $n = 0$ and $n = 1$ leads to the equations

$$A + B = 1$$
$$\tfrac{1}{2}(1+\sqrt{5})A + \tfrac{1}{2}(1-\sqrt{5})B = 1$$

Solving these equations yields

$$A = \frac{1+\sqrt{5}}{2\sqrt{5}} \quad \text{and} \quad B = -\frac{1-\sqrt{5}}{2\sqrt{5}}.$$

Hence the $n$-th Fibonacci number is

$$F_n = \frac{1}{\sqrt{5}}\left[\left(\frac{1+\sqrt{5}}{2}\right)^{n+1} - \left(\frac{1-\sqrt{5}}{2}\right)^{n+1}\right]. \quad \square$$

**The characteristic equation**

In general, putting $x_n = \lambda^n$ into the recurrence relation (15.1) leads to the *characteristic equation*

(15.8) $\qquad \lambda^k - a_1\lambda^{k-1} - a_2\lambda^{k-2} - \cdots - a_{k-1}\lambda - a_k = 0.$

For each solution $\lambda$ to this equation we find that $x_n = \lambda^n$ is a solution to the given recurrence relation.

**Remarks**

1. If $\lambda_1, \lambda_2, \ldots, \lambda_k$ are the roots of (15.8) and they are *distinct*, then the *general solution* to (15.1) is
$$A_1\lambda_1^n + A_2\lambda_2^n + \cdots + A_k\lambda_k^n.$$

2. If the characteristic equation (15.8) has a repeated root, say $\lambda$ occurs with *multiplicity* $m$, then we find that
$$\lambda^n, \quad n\lambda^n, \quad n^2\lambda^n, \quad \ldots, \quad n^{m-1}\lambda^n$$
are also solutions to the recurrence relation. This is a consequence of (14.6) applied to the partial fraction expansion of the generating function.

3. In order to describe the general solution to (15.1) let us suppose that the roots of (15.8) are $\lambda_1, \lambda_2, \ldots, \lambda_r$ and that $\lambda_i$ occurs with multiplicity $m_i$. Then the general solution is
$$\sum_{i=1}^{r}\sum_{j=1}^{m_i} A_{ij}\, n^{j-1}\, \lambda_i^n,$$
where the $A_{ij}$ are arbitrary constants.

EXAMPLE 15.9 Solve $x_n = 6x_{n-1} - 12x_{n-2} + 8x_{n-3}$, where $x_0 = 0$, $x_1 = 2$ and $x_2 = 12$.

The characteristic equation is
$$\lambda^3 - 6\lambda^2 + 12\lambda - 8 = 0.$$

That is, $(\lambda - 2)^3 = 0$ and therefore $\lambda = 2$ is a root of multiplicity 3. Thus the general solution to the recurrence relation is
$$x_n = A\, 2^n + B\, n 2^n + C\, n^2 2^n.$$

Putting $n$ equal to 0, 1 and 2 we find that
$$\begin{aligned} A &= 0 \\ 2A + 2B + 2C &= 2 \\ 4A + 8B + 16C &= 12 \end{aligned}$$

Thus $A = 0$ and $B = C = \frac{1}{2}$. Therefore the solution is
$$x_n = \frac{1}{2}(n + n^2)2^n.$$

The same solution can be obtained using the generating function $G(z)$ of the $x_n$. Its closed form turns out to be
$$G(z) = \frac{2z}{1 - 6z + 12z^2 - 8z^3} = \frac{2z}{(1 - 2z)^3}$$

and the coefficient of $z^n$ can be found using formula (14.6). ∎

## Summary

After reading this chapter you should be able to

- obtain the characteristic equation of a linear recurrence relation;
- use the roots of the characteristic equation to obtain the general solution of the recurrence relation.

## Problem Set 15

1. (i) Show that $x_n = 6 \cdot 2^n - 4$ is a solution to the recurrence relation
$$x_n = 3x_{n-1} - 2x_{n-2}.$$

   (ii) Show that $x_n = (3^{n+1} - 1)/2$ is a solution to the recurrence relation
$$x_{n+1} - x_n = 3^{n+1}.$$

   (iii) Show that $x_n = n!$ is a solution to the recurrence relation
$$x_n - n(n-1)x_{n-2} = 0.$$

2. Find a recurrence relation for the sum $S_n = \sum_{i=1}^{n} i^2$.

3. Solve the following recurrence relations:
   (i) $x_n = 7x_{n-1}$, for $n \geq 1$, where $x_0 = 2$.
   (ii) $x_n = 10x_{n-1} - 25x_{n-2}$, for $n \geq 2$, where $x_0 = -1$ and $x_1 = 5$.
   (iii) $x_{n+3} - 6x_{n+2} + 11x_{n+1} - 6x_n = 0$, for $n \geq 0$, where $x_0 = 1$, $x_1 = 0$ and $x_2 = -1$.

4. Solve the following recurrence relations:
   (i) $x_n = 4x_{n-1} - 3x_{n-2}$, where $x_0 = 1$ and $x_1 = 2$.
   (ii) $x_n = 3x_{n-1} - 3x_{n-2} + x_{n-3}$, where $x_0 = 0$, $x_1 = 1$ and $x_2 = 3$.

5. Solve the following recurrence relations:
   (i) $x_n = 5x_{n-1} - 6x_{n-2}$, where $x_0 = 1$ and $x_1 = 1$.
   (ii) $x_n - 5x_{n-1} + 8x_{n-2} - 4x_{n-3} = 0$, where $x_0 = 0$, $x_1 = 2$ and $x_2 = 4$.

# 16

# Formal Languages

FOR the remainder of the book we are concerned with the production, recognition and counting of strings of symbols.

EXAMPLE 16.1 If $\Sigma = \{a, b, c\}$, then some of the strings obtained from $\Sigma$ are *abc, aaaaaa, bacaabbca*, and *bcabcbacc*. ❑

Collections of strings of symbols are called *formal languages*. More precisely, we begin with a finite set $\Sigma$, called the *alphabet* and let $\Sigma^*$ be the set of all strings that can be made from the elements of $\Sigma$. Any subset of $\Sigma^*$ is called a *formal language*.

*ε has length 0.*   The set $\Sigma^*$ includes the *empty* string $\varepsilon$ which has no symbols in it. We need $\varepsilon$ in our theory in the same way that we need 0 in arithmetic and $\emptyset$ in set theory.

EXAMPLE 16.2 The collection of all balanced strings of brackets is a language. In this case the alphabet is the set $\{\,(,)\,\}$. Here, and in many other cases, we will be interested in finding out how many strings there are of length $n$ for each $n = 0, 1, 2, \ldots$ . ❑

EXAMPLE 16.3 In Chapter 11 we considered compound propositions constructed from the symbols $p$, $q$, $r$, ... together with brackets and the connectives $\wedge$, $\vee$, $\sim$, $\Rightarrow$ and $\Leftrightarrow$. Although we didn't discuss it there, not all strings of these symbols represent meaningful propositions. For example, $\Rightarrow p \wedge \wedge$ makes no sense. So we need some rules to tell us which strings represent compound propositions. The strings of symbols built up using these rules are called *well-formed formulae* (WFF's). Here are the rules:

($i$)   Individual variables $p, q, r, \ldots$ are WFF's.
($ii$)  If $x$ and $y$ are WFF's, so are
   (a)   $(x \vee y)$,
   (b)   $(x \wedge y)$,
   (c)   $(x \Rightarrow y)$,
   (d)   $(x \Leftrightarrow y)$,
   (e)   $\sim x$.

The collection of all **WWF**'s formed in this way is the *language of propositional calculus.* ❏

## Operations on languages

There are several important ways to create new languages out of old ones.

### Union

If $L_1$ and $L_2$ are languages we can form their union $L = L_1 \cup L_2$. That is $L$ consists of the strings that are either in $L_1$ **or** in $L_2$.

### Concatenation

A more interesting operation is *concatenation*. If $u$ and $v$ are strings, then the *concatenation* of $u$ with $v$ is the string $uv$ produced by writing $u$ followed by $v$. For example, if $u = aabcb$ and $v = cbbaab$, then $uv = aabcbcbbaab$. We can also say that $u$ and $v$ are *juxtaposed*. Note that the concatenation of the empty string $\varepsilon$ with any string produces the same string: $\varepsilon v = v\varepsilon = v$.

The idea of concatenation can be extended to languages in a natural way. If $L_1$ and $L_2$ are languages, then we define the concatenation $L_1 L_2$ of $L_1$ with $L_2$ to be

Note that $\emptyset L = \emptyset$ and $\emptyset \cup L = L$.

$$L_1 L_2 = \{\, uv \mid u \in L_1, v \in L_2 \,\}.$$

Powers of $L$ are used to denote the concatenation of a language $L$ with itself. Thus $L^2 = LL$, $L^3 = LLL$, etc. We also set $L^0 = \{\varepsilon\}$ and $L^1 = L$.

EXAMPLE 16.4 Let $\Sigma = \{a, b, c, 0\}$, and let $L_1 = \{a0, b0, c0\}$ and $L_2 = \{aa, bb\}$ be two languages. Then the concatenation $L_1 L_2$ of $L_1$ and $L_2$ is

$$L_1 L_2 = \{a0aa, b0aa, c0aa, a0bb, b0bb, c0bb\},$$

and the concatenation $L_1^2$ of $L_1$ and $L_1$ is

$$L_1^2 = \{a0a0, a0b0, a0c0, b0a0, b0b0, b0c0, c0a0, c0b0, c0c0\}. \quad ❏$$

### *-closure

Stephen C. Kleene
(1909– )

Another useful operation on languages is *-*closure* (also called the *Kleene closure*). If $L$ is any language, then $L^*$ is the set of all strings (including the empty string $\varepsilon$) that can be formed by concatenating any number of strings of $L$. That is,

$$L^* = \bigcup_{i=0}^{\infty} L^i.$$

EXAMPLE 16.5 Let $\Sigma = \{0, 1\}$ and $L = \{0, 01\}$. Then the *-closure $L^*$ is the collection of all strings of 0's and 1's in which every 1 is preceded by at least one 0. ❏

**Theorem 16.6** *If $L$ and $M$ are two languages over the same alphabet $\Sigma$ and if $L \subseteq M$, then $L^* \subseteq M^*$.*

**Proof.** This is clear because if $\alpha \in L^*$, then $\alpha = \ell_1 \ell_2 \ldots \ell_m$ is constructed by concatenating elements $\ell_1, \ell_2, \ldots, \ell_m$ of $L$. As these are also elements of $M$ (by hypothesis), we have $\alpha \in M^*$. □

# Regular expressions; regular languages

According to our definition a language can be any collection of strings whatsoever. This is too general. We now restrict ourselves to some special types of languages called *regular languages*.

A regular language will be defined by something called a *regular expression*.

A regular expression over an alphabet $\Sigma$ is a special sort of string built from the elements of $\Sigma$ and five additional symbols:

A single regular expression $r$ (to be defined below) designates a *regular language $L(r)$* over the alphabet $\Sigma$.

Regular expressions are built up from simpler regular expressions as follows.

(*i*)   $\emptyset$ is a regular expression which designates the empty language and $\varepsilon$ is a regular expression which designates the language $\{\varepsilon\}$.

(*ii*)   Each element $a$ of $\Sigma$ is a regular expression and the regular language that it designates is $\{a\}$.

(*iii*)   If $r_1$ is a regular expression that designates $L_1$ and if $r_2$ is a regular expression that designates $L_2$, then $(r_1 + r_2)$ is a regular expression that designates the language $L_1 \cup L_2$.

(*iv*)   If $r_1$ is a regular expression that designates $L_1$ and if $r_2$ is a regular expression that designates $L_2$, then $(r_1 r_2)$ is a regular expression that designates the language $L_1 L_2$, the concatenation of $L_1$ and $L_2$.

(*v*)   If $r$ is a regular expression that designates $L$, then $(r)^*$ is a regular expression that designates the language $L^*$, the $*$-closure of $L$.

As in arithmetic, brackets are omitted whenever it is convenient to do so. This is possible because $+$ and concatenation are associative and we may take $*$ to have higher precedence than concatenation and $+$, and concatenation to have higher precedence than $+$. Associativity of concatenation means that $(r_1(r_2 r_3))$ and $((r_1 r_2)r_3)$ designate the same language. To say that concatenation has higher precedence than $+$ means that we interpret $r_1 r_2 + r_3$ as $((r_1 r_2) + r_3)$ rather than $(r_1 (r_2 + r_3))$.

EXAMPLE 16.7 We can write $ab^* + a$ instead of $((a(b^*)) + a)$. ❏

Regular expressions give us an easy way to describe many languages.

EXAMPLE 16.8 The following table gives some regular expressions over the alphabet $\{a, b\}$ together with their designated languages.

| Regular expression | Designated language |
|---|---|
| $a$ | $\{a\}$ |
| $a^*$ | $\{a\}^* = \{\varepsilon, a, aa, aaa, \ldots\}$ |
| $ab$ | $\{ab\}$ |
| $a + b$ | $\{a, b\}$ |
| $(ab)^*$ | $\{ab\}^* = \{\varepsilon, ab, abab, ababab, \ldots\}$ |
| $ab^*$ | $\{a\}\{b\}^* = \{a, ab, abb, abbb, \ldots\}$ |
| $a^* + (ab)^*$ | $\{a\}^* \cup \{ab\}^* = \{\varepsilon, a, aa, \ldots, ab, abab, \ldots\}$ |
| $(ab)^*a$ | $\{ab\}^*\{a\} = \{a, aba, ababa, ababab a, \ldots\}$ |
| $(ab)^*ab^*$ | Strings that begin with any number (possibly zero) of pairs $ab$ followed by a single $a$, followed by any number (possibly zero) of $b$'s. ❏ |

Not all languages are regular. For example, the language whose strings are all strings of balanced brackets is not regular. We shall see why in Chapter 17.

## Equivalent expressions

It is possible for the same language to be designated by different regular expressions.

EXAMPLE 16.9 The regular expressions $(a+b+c)^*$ and $(a^*b^*c^*)^*$ designate the same language.

To see this, let $\Sigma = \{a, b, c\}$. Then the language designated by $(a + b + c)^*$ is $\Sigma^*$, which is the set of all possible strings that can be made from $a$, $b$ and $c$. On the other hand, $\Sigma$ is a subset of the language designated by $a^*b^*c^*$ and so by Theorem 16.6 $\Sigma^*$ is a subset of the language designated by $(a^*b^*c^*)^*$ which in turn is a subset of $\Sigma^*$. It follows that the languages are equal and therefore both expressions designate the language of all possible strings. ❏

EXAMPLE 16.10 Let $L$ be the language consisting of all strings of 0's and 1's in which every string ends in 1 and in which every 1 except the last is followed by a 0. Then $L$ can be designated by $0^*1(00^*1)^*$ and by $(0+10)^*1$. ❏

**The algebra of regular expressions**

If we agree to write $r = s$ whenever the regular expressions $r$ and $s$ designate the same language, then we have:

$$r + s = s + r$$
$$r + (s + t) = (r + s) + t$$
$$r\varepsilon = \varepsilon r = r$$
$$r(s + t) = rs + rt$$
$$(r + s)t = rt + st$$
$$(r + s)^* = (r^*s)^*r^*$$
$$(rs)^* = \varepsilon + r(sr)^*s$$
$$(r^*)^* = r^*$$

## Connections with counting

We often want to know how many strings of length $n$ there are in a given language $L$.

EXAMPLE 16.11 Let the alphabet $\Sigma = \{a, b, c, d\}$. How many strings of length 10 are there in the language designated by the regular expression $a^*b^*c^*d^*$?

This is equivalent to choosing 10 things from 4 things, allowing repetition. Therefore the answer is

$$\binom{10 + 4 - 1}{4 - 1} = \binom{13}{3} = 286. \quad ❏$$

It is sometimes possible to count the number of strings of length $n$ in a language by using recurrence relations or generating functions.

EXAMPLE 16.12 (Fibonacci numbers again) Let $L$ be the language designated by $(0 + 10)^*1$. Then $L$ consists of those strings which end with a 1 and in which every other 1 is followed by a 0. The first few strings of $L$ are:

$$1, \ 01, \ 001, 101, \ 0001, 0101, 1001$$

Let $\ell_n$ be the number of strings in $L$ of length $n$. If a string of $L$ begins with 0 it must have the form $0u$, where $u \in L$. If it begins with 1, it is either 1 itself or it has the form $10v$, where $v \in L$. Thus, for $n \geq 2$ we find that

$$\ell_n = \ell_{n-1} + \ell_{n-2}.$$

This is the recurrence relation which is satisfied by the Fibonacci numbers $F_n$ (cf. Problem 15.2) but we have the initial conditions $\ell_1 = 1$, $\ell_2 = 1$. Therefore $\ell_n = F_{n-1}$. We shall discuss this example again in Chapter 18. ❑

## Summary

After reading this chapter you should know the meaning of

- union, concatenation and $*$-closure of languages;
- regular languages;
- regular expressions;
- the language designated by a regular expression.

## Problem Set 16

1. Let $\Sigma$ be the alphabet $\{a, b, c, d, e\}$. What is $\Sigma^2$, $\Sigma^3$? What are their sizes? How many strings of length 5 are there in $\Sigma^*$?

2. If $\Sigma = \{v, w, x, y, z\}$, how many strings of length 6 in $\Sigma^*$ begin with $xy$?

3. Let $\Sigma$ be the alphabet $\{a, b, c, d, e\}$. Let $L_1 = \{ab, ac, de\}$ and $L_2 = \{bc, cd\}$ be two languages. Write down the languages $L_1 L_2$ and $L_2 L_1$.

4. For each of the following regular expressions over the alphabet $\Sigma = \{a, b, c, d\}$, describe the designated language.
   - (i) $a + bc + cd$
   - (ii) $b + cd^*$
   - (iii) $(ab + cd)^*$
   - (iv) $(dd)^*$
   - (v) $(ab)^*(db)^*$
   - (vi) $(ddd)^*$

5. Show that each of the following languages over $\Sigma = \{a, b, c\}$, is regular by finding a regular expression which designates it.
   - (i) $\{\varepsilon, a, b, b^2, b^3, \ldots\}$,
   - (ii) $\{abc, abcabc, abcabcabc, \ldots\}$,
   - (iii) $\{b, ab, a^2b, a^3b, \ldots\}$,
   - (iv) $\{\varepsilon, a, b, c, ab, abab, ababab, \ldots\}$.

# 17

# Finite State Machines

IN this chapter we consider certain very simple machines which are capable of recognizing the strings of a regular language. That is, for each regular language we want to construct a machine that can somehow read a string (from left to right) and then tell us whether or not that string is in the language.

*The singular of automata is automaton.*

These machines have a finite number of *states* and are called *finite automata* or *finite state machines*.

## Deterministic finite automata

We begin with a particularly simple type of machine called a *deterministic finite automaton* or DFA for short. We can draw such a machine using circles to indicate the states and arrows labelled with letters from some alphabet to indicate transitions from state to state. Some states, called *accepting states*, are represented by double circles. For example, Figure 17.1 is a DFA with 5 states:

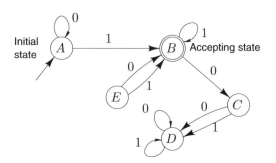

Figure 17.1
A finite state machine

See Example 17.4 for more details.

There is also an unlabelled arrow indicating the state at which the machine begins, called the *initial state*. The idea is that the machine begins in the initial state and reads a string of 0's and 1's (from its left hand end) and changes state according to its current state and the symbol just read. If the machine ends up in an accepting state, it *accepts* the string, otherwise it *rejects* it.

The precise definition of a *deterministic finite automaton* or DFA is that it consists of:

Also called an input alphabet.

(*i*) An *alphabet* $\Sigma$.
(*ii*) A finite set $\mathcal{S}$ of states.
(*iii*) An *initial state*.
(*iv*) A set $\mathcal{A} \subseteq \mathcal{S}$ of *accepting* states.
(*v*) A *transition function* $f : \mathcal{S} \times \Sigma \to \mathcal{S}$.

There is exactly one arrow labelled $x$ leaving $A$.

We draw the diagram of a DFA by drawing a circle for each state, and we use double circles to mark the accepting states. Then for each state $A \in \mathcal{S}$ and each symbol $x \in \Sigma$ we draw an arrow labelled $x$ from $A$ to $f(A, x)$. The initial state is marked by an unlabelled arrow leading to it.

The machine is called *deterministic* because given any state $A$ and any symbol $x$, the next state is completely determined by $A$ and $x$.

For our convenience, in drawing diagrams we sometimes label an arrow with more than one symbol. This is an abbreviation for several arrows, each with a single symbol.

**Dead-end states**

A *dead-end* state is one from which it is impossible to get to an accepting state, no matter what the input. Sometimes we omit dead-end states from our diagrams. In Figure 17.1, $C$ and $D$ are dead-end states. After we remove dead-end states there is at most one arrow with a given label leaving each state.

**Inaccessible states**

An *inaccessible* state is one which cannot be reached by following arrows from the initial state. As with dead-end states, it is often convenient to omit them from our diagrams. In Figure 17.1, $E$ is an inaccessible state.

**Paths**

Let $M$ be a DFA. Then a *path* in the diagram of $M$ is a sequence of arrows

$$a_0 \, a_1 \, \ldots \, a_k$$

such that the state that $a_{i-1}$ leads to is the same state that $a_i$ leaves from, for $i = 1, 2, \ldots, k$. If arrow $a_i$ carries the label $x_i$, then the string $x_0 \, x_1 \, \ldots \, x_k$ is the label of the path.

EXAMPLE 17.1 In Figure 17.1 there is a path from $A$ to $D$ with label 11100. It goes from $A$ to $B$, to $B$, to $B$, to $C$, then to $D$.

**Accepted strings**

A string is *accepted* by $M$ if there is a path beginning at the initial state and ending at an accepting state whose arrows are labelled by the symbols in the string.

EXAMPLE 17.2 Consider the DFA $M$ consisting of
(i) The alphabet $\Sigma = \{a, b, c\}$.
(ii) The finite set $\mathcal{S} = \{A, B, C, D, E\}$ of states.
(iii) The initial state $A$.
(iv) The set $\mathcal{A} = \{C, D\} \subseteq \mathcal{S}$ of *accepting* states.
(v) The transition function given by the following table in which the labels on the rows are the states, the labels on the columns are the input symbols and the entries indicate the next state.

|   | a | b | c |
|---|---|---|---|
| A | B | E | E |
| B | C | D | E |
| C | E | E | E |
| D | E | D | E |
| E | E | E | E |

The DFA $M$ can be drawn as in Figure 17.2.

Figure 17.2

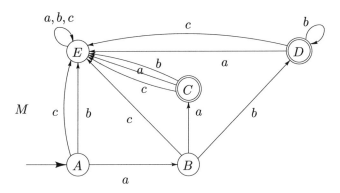

Notice that by omitting the dead-end state $E$, the diagram can be drawn more simply as:

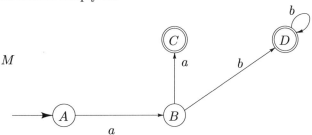

The state $E$ can be omitted from the transition function and it can be written more compactly as:

|   | $a$ | $b$ |
|---|---|---|
| $A$ | $B$ | – |
| $B$ | $C$ | $D$ |
| $C$ | – | – |
| $D$ | – | $D$ |

Now the label $c$ no longer appears and this makes it clear that no string containing $c$ can be accepted by the machine $M$. It is easy to see that $M$ accepts only the strings $aa$, $ab$, $abb$, ... . That is, either $a$ followed by a single $a$ or by a string of at least one $b$'s. ❑

## The language of a machine

The strings accepted by a DFA $M$ form a language denoted by $L(M)$. We call this the language *accepted* by $M$.

EXAMPLE 17.3 For the DFA $M$ of Figure 17.2, the language $L(M)$ is

$$L(M) = \{aa, ab, abb, abbb, \ldots\},$$

which is designated by the regular expression $aa + abb^*$.

EXAMPLE 17.4 In order to find the language accepted by the DFA of Figure 17.1 we first remove the dead-end and inaccessible states then redraw the diagram:

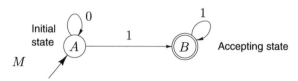

The input alphabet is $\{0, 1\}$ and it is clear that to get from the initial state to the accepting state, the arrow from $A$ to $B$ must be used at some point. Before this you can go around the loop at $A$ as many times as you like and after you get to $B$ you can go around the loop there as many times as you like. Thus the language $L(M)$ accepted by $M$ consists of strings of any number of 0's followed by at least one 1. This can be written $\{0\}^*\{1\}\{1\}^*$, which is designated by the regular expression $0^*11^*$.

## Kleene's theorem

In the previous two examples we made the observation that the language $L(M)$ of the DFA could be described by a regular expression. As shown by the following theorem, this is no accident.

**Theorem 17.5**

*We sketch a proof of this below.*

(i) The language $L(M)$ accepted by a DFA is regular. That is, it can be designated by a regular expression in the input alphabet.

(ii) For every regular language $L$ there is some DFA $M$ such that $L(M) = L$.

This says that the regular languages are precisely those languages which can be recognized by deterministic finite automata.

EXAMPLE 17.6 Given a DFA, it is sometimes possible to find a regular expression which describes the language it accepts simply "by inspection". We do this by first listing a few representative strings accepted by the machine and then look for a common pattern.

*This only works for fairly simple machines.*

Figure 17.3

*Note that D is a dead-end state.*

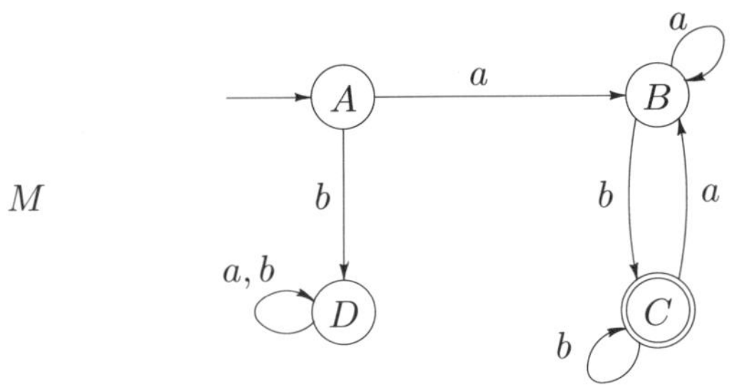

The machine $M$ of Figure 17.3 accepts the strings

$$ab, \quad abab, \quad ababb$$
$$aab, \quad aaabababb, \quad abaaababbb.$$

After making this list we are led to conjecture that the language accepted by $M$ consists of all strings which begin with $a$ and end with $b$. We can check directly that this is true. That is,

$$L(M) = \{\, aub \mid u \text{ is any string of } a\text{'s and } b\text{'s}\,\}$$
$$= \{a\}\{a, b\}^*\{b\},$$

and so $L(M)$ is designated by the regular expression $a(a+b)^*b$.
□

**The regular expression of a DFA**

We now fulfill our promise to sketch a proof of the first part of Kleene's theorem. (The second part is left as a rather challenging exercise.)

Let $S$ be the set of states of a DFA $M$. For each pair of states $A$, $B$ and each subset $X$ of $S$ we let $L(A, B, X)$ be the language consisting of the labels of all paths which start at $A$ and get to $B$ via states in $X$. If there are no arrows from $A$ to itself we define $L(A, A, \emptyset)$ to be $\{\varepsilon\}$. Note that we don't require $A$ or $B$ to belong to $X$.

We shall prove by induction on $|X|$ that for all $X$ there exists a regular expression $\rho(A, B, X)$ which defines $L(A, B, X)$. Suppose at first that $|X| = 0$, i.e., $X = \emptyset$. If there are no arrows from $A$ to $B$, then by definition $L(A, B, \emptyset)$ is $\{\varepsilon\}$ or $\emptyset$ according to whether $A = B$ or $A \neq B$. If the arrows from $A$ to $B$ are labelled $x_1, x_2, \ldots, x_k$, we take $\rho(A, B, \emptyset)$ to be $x_1 + x_2 + \cdots + x_k$.

Now suppose that $|X| > 0$. Choose $C \in X$ and put $Y = X \setminus \{C\}$. By induction we can find regular expressions $\rho(A, B, Y)$, $\rho(A, C, Y)$, $\rho(C, C, Y)$ and $\rho(C, B, Y)$ which define the languages $L(A, B, Y)$, $L(A, C, Y)$, $L(C, C, Y)$ and $L(C, B, Y)$, respectively. We now put

$$\rho(A, B, X) = \rho(A, B, Y) + \rho(A, C, Y)\rho(C, C, Y)^*\rho(C, B, Y).$$

We claim that this is a regular expression which defines $L(A, B, X)$. It is a regular expression simply because it is built up from other regular expressions using the operations of addition, concatenation and $*$-closure. To see that it defines the language $L(A, B, X)$, note that for any path that goes from $A$ to $B$ via states in $X$ there are two possibilities: either it goes from $A$ to $B$ without using $C$, in which case its label is in $L(A, B, Y)$; or else it uses $C$ at least once, in which case its label is in the language

*See Chapter 16.*

$$L(A, C, Y)L(C, C, Y)^*L(C, B, Y).$$

This completes the induction step.

If $I$ is the initial state, then the regular expression which defines $L(M)$ is the sum of the regular expressions $\rho(I, A, S)$, where $A$ ranges over the accepting states. □

## Non-deterministic finite automata

A DFA is *deterministic* in the sense that for each state and each input symbol, the next state is completely determined. We now consider a generalization of the idea of a deterministic finite automaton, namely a *non-deterministic finite automaton* or NFA for short. The following diagram depicts an NFA. (Dead-end states have been omitted.) In an NFA, arrows can be labelled with the empty string $\varepsilon$.

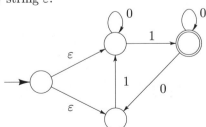

*There can be several arrows with the same label starting at a state.*

More precisely, an NFA consists of

(*i*) An *alphabet* $\Sigma$.

*Some books allow more than one initial state.*

(*ii*) A finite set $\mathcal{S}$ of states.

(*iii*) An *initial* state.

(*iv*) A set $\mathcal{A} \subseteq \mathcal{S}$ of *accepting* states.

$\varepsilon$ *is the empty string.*

(*v*) A *transition function* $f : \mathcal{S} \times (\Sigma \cup \{\varepsilon\}) \to \mathcal{P}(\mathcal{S})$.

This is almost the same description as for a DFA except that the transition function takes a state and an input symbol to a

*Recall that $\mathcal{P}(X)$ denotes the set of subsets of the set $X$.*

subset of the set of states.

It is perhaps easier to think of an NFA in terms of its diagram. The diagram is constructed in exactly the same way as for a DFA except that each state may have more than one arrow leaving it with the same label. Also, we are allowed to use the empty string $\varepsilon$ to label arrows. Then $f(Z, x)$ is the set of all states $W$ such that there is an arrow from $Z$ to $W$ labelled $x$.

**The language of an NFA.**

Let $M$ be an NFA. To each path $p$ of arrows in the diagram of $M$ we associate the string $u(p)$ obtained by concatenating the symbols along $p$. We say that a string $v$ is *accepted* by $M$ if there is some path $p$ from the initial state to one of the accepting states such that $u(p) = v$. The set of all strings accepted by $M$ is the language $L(M)$ accepted by $M$.

It turns out that the language accepted by an NFA is regular and so by Kleene's theorem there is always a DFA that accepts the same language. But if the NFA has $m$ states then the DFA may have $2^m$ states.

On the other hand if we begin with a regular language it is often much easier to find an NFA that accepts it, rather than a DFA.

# Non-regular languages

Kleene's theorem states that a language is regular if and only if it can be recognized by a DFA. We can use this to show that certain languages are *not* regular.

EXAMPLE 17.7 Let $L$ be the language consisting of all strings of $n$ 0's followed by $n$ 1's for $n = 0, 1, 2, \ldots$. We claim that $L$ is not regular.

Suppose that $L$ could be recognized by a DFA $M$ with $m$ states. Then consider the string

$$\underbrace{00\cdots 0}_{m}\underbrace{11\cdots 1}_{m}.$$

*This uses the Pigeonhole Principle.*

We know that there should be a path from the initial state to an accepting state which is labelled by this string. But $M$ has only $m$ states, and therefore the first part of the string (corresponding to the $m$ 0's) must go through the same state, say $K$, at least twice. Let $u$ be the path from the first to the second visit of $K$. This path will be labelled by $k$ 0's, where $k$ is the number of arrows in the path (and hence $k$ will be at least 1). But $u$ starts at $K$ and ends at $K$, and so we can go around it as many times as we like and then continue on as before. This means that the machine must also accept a string beginning with $m + k$ 0's followed by $m$ 1's. But this string is not in $L$. This contradiction means that we cannot construct a DFA which recognizes only the strings of $L$. Hence $L$ is not regular. ❑

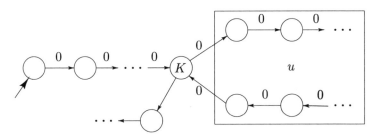

EXAMPLE 17.8 There is no DFA which recognizes only balanced strings of brackets. This is because we can apply the above argument to the strings of the form

$$\underbrace{((\cdots(}_{m}\underbrace{))\cdots)}_{m}.$$ ❑

The argument can be generalized further. If $L$ is the language accepted by a DFA with $m$ states, then every string of $L$ of length at least $m$ can be written in the form $uvw$, where the length of $uv$ is at most $m$, $v \neq \varepsilon$ and for $i = 0, 1, 2, \ldots$, the string $uv^i w$ is in $L$.

## Summary

After reading this chapter you should know the meaning of

- DFA, NFA;
- initial state, accepting state;
- transition function;
- the language of a DFA or NFA;
- Kleene's theorem.

# Problem Set 17

1.

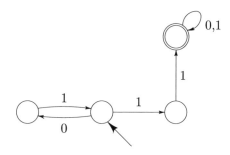

(a) Which of the following strings are accepted by the DFA which appears above?

(i)   001101      (ii)  0001      (iii) 11
(iv)  110100      (v)   1111      (vi)  00100

(b) For each regular expression below, state whether or not it designates the language accepted by the finite state machine.

(i)   $(0+1)^*11(0+1)^*$
(ii)  $(01)^*11(0+1)^*$
(iii) $(01)^*11(01)^*$

2.

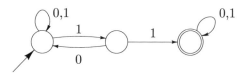

(a) Which of the following strings are accepted by the NFA which appears above?

(i)   001101      (ii)  0001      (iii) 11
(iv)  110100      (v)   1111      (vi)  00100

(b) For each regular expression, state whether or not it designates the language accepted by the finite state machine.

(i)   $0^*(0+1)1(0+1)^*$
(ii)  $0^*1(00^*1)^*(0+1)^*$
(iii) $(0+10)^*11(0+1)^*$

3. (i) Draw the DFA that corresponds to the following transition function. The alphabet is $\{a, b\}$, the initial state is $A$ and the only accepting state is $B$.

|   | a | b |
|---|---|---|
| A | B | C |
| B | B | C |
| C | B | D |
| D | D | D |

(ii) Which of the following strings are accepted by the DFA given in (i)?

(a)  aaa  (b)  aaab
(c)  bbab  (d)  baba

(iii) Describe the language accepted by the machine in (i).

(iv) Find a regular expression which designates the language found in (iii).

4. Construct a DFA that accepts only those strings of lower case letters which end in "ing".

*5. Given regular expressions $r_1$ and $r_2$ and NFA's $M_1$ and $M_2$ which accept the languages $L(r_1)$ and $L(r_2)$, describe how to construct new NFA's (using $M_1$ and $M_2$) which accept the languages

(i) $L(r_1 + r_2)$,   (ii) $L(r_1 r_2)$,   (iii) $L(r_1^*)$.

# 18

# Grammars

IN the last chapter we mentioned that the language accepted by a deterministic finite automaton is regular. The regular languages are very special and indeed we have seen that the language of balanced strings of brackets is not regular. In this chapter we study a method of describing a broader class of languages. To do this we introduce the idea of a *grammar*.

## Grammars

To describe a *grammar* we need two alphabets and a collection of rules known as *productions*. We let

(i)  $T$ be the alphabet of *terminal* symbols,
(ii) $N$ be the alphabet of *non-terminal* symbols, and we assume that $T$ and $N$ have no symbols in common,
(iii) $S \in N$ be the *start* symbol,
(iv) $P$ be the set of *productions* of the form $\alpha \to \beta$ where $\alpha$ and $\beta$ are strings constructed from the terminal and nonterminal symbols. The string $\alpha$ must contain at least one nonterminal symbol.

Thus a *grammar* can be thought of as a 4-tuple
$$G = (N, T, S, P).$$

## The language of a grammar

From the grammar $G$ we can produce strings in the terminal alphabet $T$. Then the language, $L(G)$, produced by the grammar $G$ is the collection of strings that can be produced from the start symbol by a sequence of productions.

The application of a sequence of productions is called a *derivation*.

**Problem 18.1** Given a grammar $G$, find the language $L(G)$ produced by $G$.

**Solution.** The alphabet $T$ of terminal symbols of the grammar is the alphabet of the language $L(G)$. We produce the strings in

$L(\mathcal{G})$ as follows. We begin by writing down $S$. At any stage we may replace a symbol which occurs to the left of the arrow of a production by the string to the right of that arrow. We keep going until only terminal symbols remain in the string. ❑

EXAMPLE 18.2 Consider the grammar $\mathcal{G} = (\mathcal{N}, \mathcal{T}, S, \mathcal{P})$ where $\mathcal{N} = \{S, A, B\}$, $\mathcal{T} = \{0, 1\}$ and the productions are

$$S \to ASB$$
$$S \to AB$$
$$A \to 0$$
$$B \to 1.$$

We can derive the string 0011 by the following productions:

$$S \to ASB$$
$$\to AABB$$
$$\to 0ABB$$
$$\to 00BB$$
$$\to 001B$$
$$\to 0011$$

In fact, the language $L(\mathcal{G})$ generated by $\mathcal{G}$ consists all the strings of the form

$$\underbrace{00\cdots0}_{m}\underbrace{11\cdots1}_{m} \qquad m \geq 1$$

That is,
$$L(\mathcal{G}) = \{01,\, 0011,\, 000111,\, \ldots\}.$$

This example shows that the language produced by a grammar *need not* be regular. ❑

Productions can use the empty string $\varepsilon$ and so in Example 18.2 the productions can be written more simply as

$$S \to 0A1$$
$$A \to 0A1$$
$$A \to \varepsilon.$$

Note that the empty string is not in the language. Compare this with the next example.

EXAMPLE 18.3 Let $\mathcal{G}_1 = (\mathcal{N}, \mathcal{T}, S, \mathcal{P}_1)$ be a grammar, where $\mathcal{N} = \{S, A, B\}$, $\mathcal{T} = \{0, 1\}$ and the productions are

$$S \to 0S1$$
$$S \to \varepsilon.$$

Then it is easy to see that the language produced by $\mathcal{G}_1$ is

$$L(\mathcal{G}_1) = \{\varepsilon,\ 01,\ 0011,\ 000111,\ \ldots\}.$$

The only difference between $L(\mathcal{G})$ and $L(\mathcal{G}_1)$ is that $L(\mathcal{G}_1)$ contains the empty string $\varepsilon$. ❑

EXAMPLE 18.4 Let $\mathcal{G} = (\mathcal{N}, \mathcal{T}, S, \mathcal{P})$ be a grammar, where $\mathcal{N} = \{S, A, B\}$, $\mathcal{T} = \{a, b\}$ and the productions are

$$S \to AB$$
$$B \to bB$$
$$A \to a$$
$$B \to b.$$

Then we can derive the string *abbb* as follows:

$$S \to AB$$
$$\to AbB$$
$$\to abB$$
$$\to abbB$$
$$\to abbb.$$

In fact, the language produced by the grammar $\mathcal{G}$ is

$$L(\mathcal{G}) = \{ab,\ abb,\ abbb,\ \ldots\} = \{ab\}\{b\}^*,$$

which is designated by the regular expression $abb^*$. ❑

## A grammar for a language

In general it is a difficult problem to find a grammar that produces a given language. In some cases, such as natural languages like English, Chinese, French or German, it is not even clear what constitutes the language. On the other hand, computer languages are generally easier to deal with and grammars are routinely used to construct compilers and interpreters for such languages.

The best we can do at this stage is to consider a simple example.

## GRAMMARS

EXAMPLE 18.5 Let $L$ be the language

$$L = \{aaa\}\{aaa\}^*.$$

We shall find a grammar for $L$.

The language consists of all strings of a multiple of 3 $a$'s: $aaa$, $aaaaaa$, .... So our aim is to produce the strings $aaa$, $aaaaaa$, $aaaaaaaaa$, .... We could use the start symbol $S$ to produce the string $aaa$ ($S \to aaa$) and then use the production $S \to aaaS$ to produce any number of $aaa$'s. Hence a grammar for $L$ is $\mathcal{G} = (\mathcal{N}, \mathcal{T}, S, \mathcal{P})$, where $\mathcal{N} = \{S\}$, $\mathcal{T} = \{a\}$ and the productions are

$$S \to aaa$$
$$S \to aaaS.\quad\square$$

The same language can be produced by many different grammars.

EXAMPLE 18.6 The language $L$ of Example 18.5 can also be produced by the grammar $\mathcal{G} = (\mathcal{N}, \mathcal{T}, S, \mathcal{P})$, where $\mathcal{N} = \{S, A, B, C\}$, $\mathcal{T} = \{a\}$ and the productions are

$$S \to aA$$
$$A \to aB$$
$$B \to aC$$
$$C \to aA$$
$$C \to \varepsilon.\quad\square$$

**Remarks**

1. It turns out that any regular language can be produced by a grammar with the following simple structure. The only productions are of the form $\alpha \to a\beta$ or $\alpha \to a$, where $\alpha$ and $\beta$ are single non-terminal symbols and $a$ is either a terminal symbol or $\varepsilon$. Conversely, the language produced by such a grammar is regular.
2. The grammars in which all the productions are of the form $\alpha \to \beta$, where $\alpha$ is a single non-terminal symbol, are called *context-free* grammars.

**Fibonacci numbers again**

Let $L$ be the language designated by the regular expression

$$(0+10)^*1,$$

and let $\ell_n$ be the number of strings of length $n$ in $L$. We have shown in Example 16.12 that $\ell_n$ is just the Fibonacci number

$F_{n-1}$ and we can see that $L$ is exactly the language accepted by the DFA:

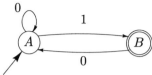

Recall that the strings in $L$ can be described as follows. Every string of $L$ that begins with 0 must have the form $0u$, where $u \in L$, and every string of $L$ that begins with 1 is either 1 itself or has the form $10v$, where $v \in L$. The description of the strings of $L$ just given shows that $L$ is the language produced by the grammar $\mathcal{G} = (\mathcal{N}, \mathcal{T}, S, \mathcal{P})$, where $\mathcal{N} = \{S\}$, $\mathcal{T} = \{0, 1\}$ and the productions are

$$S \to 0S$$
$$S \to 10S$$
$$S \to 1.$$

## Summary

After reading this chapter you should know

- the definition of a grammar;
- the connection between languages and grammars;
- that the language of a grammar need not be regular.

## Problem Set 18

1. Describe the language generated by the following grammars. The start symbol is $S$.

   (i) The non-terminal symbols are $S$, $A$ and $B$, the terminal symbols are 0 and 1 and the productions are

   $$S \to ABS$$
   $$A \to 0$$
   $$B \to 1$$
   $$S \to \varepsilon.$$

   (ii) The non-terminal symbol is $S$, the terminal symbol is $a$ and the productions are

   $$S \to aaaS$$
   $$S \to \varepsilon.$$

2. Give a grammar that has the given set of strings as its language.

   (i) All strings over $\Sigma = \{p\}$ consisting of an even number of $p$'s

   (ii) All strings over $\Sigma = \{p, q, r\}$ consisting of $n$ $p$'s followed by $n$ $q$'s, followed by a single $r$, where $n = 1, 2, 3 \ldots$ .

3. Consider the grammar with start symbol $S$, non-terminal symbols $S, A, B, C$, terminal symbols $a, b, c$ and productions

$$S \to Cc$$
$$C \to ACB$$
$$C \to AB$$
$$A \to a$$
$$B \to b$$

Describe the language of this grammar and show that it is not regular.

4. Consider the grammar with start symbol $S$, non-terminal symbols $S, A, B, C$, terminal symbols $a, b, c$ and productions

$$S \to Sc$$
$$S \to C$$
$$C \to ACB$$
$$C \to \varepsilon$$
$$A \to a$$
$$B \to b$$

Describe the language of this grammar and show that it is not regular.

# 19

# Graphs, Trees and Catalan Numbers

In Chapter 1 the Catalan numbers $c_n$ were defined as the number of balanced strings of $n$ pairs of brackets (Problem 1.1). The balanced strings of brackets are the strings of the language $L$ produced by the grammar $\mathcal{G} = \{\mathcal{N}, \mathcal{T}, S, \mathcal{P}\}$ where $\mathcal{T} = \{(,)\}$ is the alphabet of terminal symbols, $\mathcal{N} = \{S\}$ is the alphabet of non-terminal symbols, $S$ is the start symbol and the productions are

$$S \to S(S)$$
$$S \to \varepsilon.$$

If a string $w$ has the form $uv$, then $u$ is called a *prefix* of $w$. Notice that a string of $n$ ('s and $n$ )'s is in the language $L$ if and only if each prefix has at least as many left brackets "(" as right brackets ")".

## Trees

Here is another interpretation of the language of brackets. Consider "trees" planted in the ground and living in two dimensions. These are called *planted planar trees*. The dots in the drawing below are called *vertices*. A tree of $n$ vertices has $n - 1$ lines connecting the vertices; there are no circuits. Here are some examples:

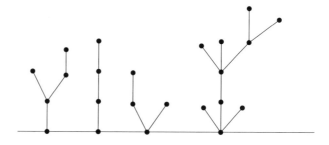

# GRAPHS, TREES AND CATALAN NUMBERS

How many planted planar trees are there with $n$ vertices? To answer this, consider a caterpillar walking across the diagram from left to right up and over the tree.

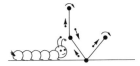

Figure 19.1

Each time the caterpillar moves up from one vertex to the next we write down "(". Each time it moves down we write ")". If the tree has $n+1$ vertices we get a string of $n$ left and $n$ right brackets in which each prefix has at least as many left brackets as right brackets. That is, the number of planted planar trees is the same as the number of balanced strings of brackets.

Actually, writing brackets in this way can get a bit confusing because we must distinguish between brackets used in the string and brackets used for the usual purpose of placing material in parentheses. To avoid this we will write $U$ (for **up**) instead of "(", and $D$ (for **down**) instead of ")". Then the caterpillar's journey, as in Figure 19.1, can be described by the string $UUDDUD$. Also, the productions for this language become

$$S \to SUSD$$
$$S \to \varepsilon.$$

We call the strings of this language *balanced* strings of $U$'s and $D$'s.

## A formula for the Catalan numbers

Let $c_n$ be the Catalan numbers. Then for each $n$,

(19.1) $$c_n = \frac{1}{n+1}\binom{2n}{n}.$$

To prove the formula, we note that there is yet another way to interpret and count the *balanced* strings of $U$'s and $D$'s.

We draw a graph corresponding to a string $u$. First we put an extra $U$ at the start of the string. This forces each prefix to have more $U$'s than $D$'s.

Now we begin at the origin $(0,0)$ and we read the string from left to right.

(i) Each time we read a $U$ we move **up** one unit and to the right one unit. That is we move from $(x, y)$ to $(x+1, y+1)$.

(*ii*) Each time we read a $D$ we move **down** one unit and to the right one unit. That is we move from $(x, y)$ to $(x+1, y-1)$.

Since we have $n+1$ $U$'s and $n$ $D$'s we must end up at the point $(2n+1, 1)$. The following figure is the graph for the string $u = UUDDUD$. Remember we first put on an extra $U$ to change this to $UUUDDUD$.

Figure 19.2

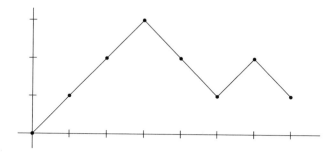

Notice that when $v$ is a balanced string of $U$'s and $D$'s, the graph of $Uv$ has the property that after leaving the origin it is always above the $x$-axis. Given *any* string of $n+1$ $U$'s and $n$ $D$'s we can draw its graph. But this graph will lie above the $x$-axis if and only if the string has the form $Uv$, where $v$ is balanced. How many strings of $n+1$ $U$'s and $n$ $D$'s are there? This is just the number of ways to choose the $n$ places to put the $D$'s, i.e.,

$$\binom{2n+1}{n}.$$

Next we need to know how many of these strings are balanced. We do this as follows. The important observation is that if we remove the symbol from the left hand end of the string and put it on the right hand end we still have a string of $n+1$ $U$'s and $n$ $D$'s but out of the $2n+1$ different strings that we get by cycling the letters around *only one is balanced.*

To see this, begin with the graph of a string $v$. This goes from $(0,0)$ to $(2n+1, 1)$ but it may touch or cross the $x$-axis. Now consider a line of slope $1/(2n+1)$ placed below this graph and raise this line until it just touches the graph. This line is parallel to the line joining the origin to the point $(2n+1, 1)$. The line can only touch the graph in one place and so the place where it touches divides the string into two pieces, say $v_1$ and $v_2$. The original string is $v_1 v_2$. But now the graph of the string $v_2 v_1$ is obtained by moving the first piece of the old graph to the right hand end. That is, we just cycle around the symbols from the left hand end to the right hand end. The graph of the new string only touches the $x$-axis at $(0, 0)$.

Figure 19.3

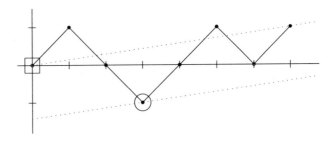

For example, the graph of the string $v = UDDUUDU$ in Figure 19.3 becomes

Figure 19.4

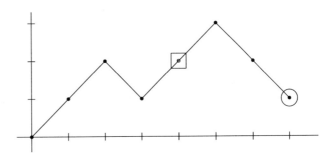

There are $2n+1$ places where the string can be broken and only one of them produces a string of the form $Uu$, where $u$ is balanced. Hence the number of balanced strings is

$$c_n = \frac{1}{2n+1}\binom{2n+1}{n}$$
$$= \frac{1}{2n+1}\frac{(2n+1)!}{(n+1)!\,n!}$$
$$= \frac{(2n)!}{(n+1)n!\,n!}$$
$$= \frac{1}{n+1}\binom{2n}{n}.$$

## A recurrence relation for Catalan numbers

Let $c_n$ be the $n$-th Catalan number. We shall prove the recurrence relation:

(19.2) $\quad c_{n+1} = c_0 c_n + c_1 c_{n-1} + c_2 c_{n-2} + \cdots + c_{n-2} c_2 + c_{n-1} c_1 + c_n c_0.$

Let $\mathcal{A}$ be the set of all balanced strings of $n$ pairs of brackets. Then as we saw above, $\mathcal{A}$ can be considered to be the set of all balanced strings of $n$ pairs of $U$'s and $D$'s. Next let $\mathcal{B}$ be the set of strings obtained by inserting $n$ pairs of brackets into the string

$$x_0 x_1 x_2 \ldots x_n$$

so that within each pair of brackets there are two terms (cf. Problem 1.7).

Before proving the recurrence relation for $c_n$, we first establish a one-to-one correspondence between $\mathcal{A}$ and $\mathcal{B}$. Suppose that $v \in \mathcal{A}$, i.e., $v$ is a balanced string of $n$ pairs of $U$'s and $D$'s. Note that $v$ always starts with a $U$ and ends with a $D$. We put an extra $U$ at the start of the string $v$ and consider the string $Uv$. Then we use the following instructions to obtain a corresponding string in $\mathcal{B}$.

Reading from left to right replace the $U$'s by $x_0$, $x_1$, $x_2$, ..., $x_n$. Then replace every $D$ by a right bracket. Again reading from left to right, each time we encounter a right bracket we place a left bracket to the left of the preceding two terms.

This construction can be reversed by removing all the left brackets, removing $x_0$, then replacing each of the $x_1$, $x_2$, ..., $x_n$ by a $U$ and replacing each right bracket by a $D$.

For example, let $v = UUDUDD \in \mathcal{A}$. Then the following illustrates the above method of obtaining an element in $\mathcal{B}$:

$$\begin{aligned} Uv = UUUDUDD &\to x_0 x_1 x_2)x_3)) \\ &\to x_0(x_1 x_2)x_3)) \\ &\to x_0((x_1 x_2)x_3)) \\ &\to (x_0((x_1 x_2)x_3)). \end{aligned}$$

Thus the above construction gives us a function $f : \mathcal{A} \to \mathcal{B}$. It is easy to see that $f$ is one-to-one. Moreover using the reverse of this construction we see that $f$ is onto. Hence $f$ is the required one-to-one correspondence between $\mathcal{A}$ and $\mathcal{B}$.

This correspondence shows that the Catalan number $c_n$ is the number of ways of bracketing

$$x_0 x_1 x_2 \ldots x_n$$

with $n$ pairs of brackets so that within each pair of brackets there are two terms.

**Proof** of (19.2): Consider the string

$$x_0 x_1 x_2 \ldots x_n x_{n+1}.$$

Then $c_{n+1}$ is the number of ways of bracketing the string with $n+1$ pairs of brackets so that within each pair of brackets there are two terms.

To obtain a bracketing of such a string, we can first bracket $x_0 x_1 x_2 \ldots x_k$, then bracket $x_{k+1} x_{k+2} \ldots x_{n+1}$, $(0 \leq k \leq n)$, and finally place a pair of brackets around the two expressions so obtained. For each $k = 0, 1, 2, \ldots, n$, the number of ways to obtain the required bracketing is $c_k c_{n-k}$. Hence, taking $c_0 = 1$, we see that

$$c_{n+1} = c_0 c_n + c_1 c_{n-1} + \cdots + c_{n-1} c_1 + c_n c_0. \quad \square$$

## The generating function for the Catalan numbers

Let $c_n$ be the $n$-th Catalan number and let

$$C(z) = \sum_{n=0}^{\infty} c_n z^n = c_0 + c_1 z + c_2 z^2 + c_3 z^3 + \cdots,$$

be the corresponding generating function. Then

(19.3) $$zC(z)^2 = C(z) - 1.$$

**Proof.** On multiplying $C(z)$ by itself, we obtain

$$C(z)^2 = c_0^2 + (c_0 c_1 + c_1 c_0)z + (c_0 c_2 + c_1 c_1 + c_2 c_0)z^2 + \cdots$$
$$= c_0^2 + c_2 z + c_3 z^2 + \cdots,$$

using (19.2). Multiply throughout by $z$ and use the fact that $c_0 = 1$, $c_1 = 1$ to obtain

$$zC(z)^2 = c_1 z + c_2 z^2 + c_3 z^3 + \cdots = C(z) - 1. \quad \square$$

By solving the quadratic equation (19.3) we obtain the following closed form for $C(z)$:

$$C(z) = \frac{1 - \sqrt{1 - 4z}}{2z}.$$

You might think about how to expand this to find the formula for $c_n$ obtained earlier.

# Summary

After reading this chapter you should know

- new ways to look at old problems (Catalan numbers);
- how generating functions and other techniques can be used to find formulae for the Catalan numbers.

## Problem Set 19

1. Prove that the Catalan numbers $c_n$ satisfy the relations:

   (i)
   $$c_n = \frac{4n-2}{n+1} c_{n-1}.$$

   (ii)
   $$c_n = \binom{2n}{n} - \binom{2n}{n+1}.$$

   (iii)
   $$c_n c_{n+1} = \sum_{k=0}^{n} \binom{2n}{2k} c_{n-k} c_k.$$

*2. Suppose that $a$ and $b$ are positive integers. Prove that the number of paths from $(0, 0)$ to $(a + b, a - b)$ which do not cross below the $x$-axis is
$$\frac{a-b+1}{a+1} \binom{a+b}{b}.$$

## 20

# Hints and Answers

**Problem Set 1**
1. $(i)$ Not balanced, $(ii)$ Balanced, $(iii)$ Not balanced.
2. $(i)$ 1, 2, 5, 14.
   $(ii)$ The arrangement 5, 1, 4, 2, 3, is possible, but not 3, 2, 4, 1, 5.
3. 1, 2, 6, 24.
4. $(i)$ 1, 1, 2, 5.  $(ii)$ The Catalan numbers.
5. 1, 2, 5, 14 for $n = 2, 3, 4, 5$. (The Catalan numbers.)
6. $(i)$ $a_2 = 2$ and $a_3 = 5$;  $(ii)$ $a_n$ is the $n$-th Catalan number.
7. 1, 2, 5, 14. (The Catalan numbers.) The numbers in the top row give the positions of the left brackets in the string. The numbers in the bottom row give the positions of the right brackets.
8. 1, 2, 5, 14, 39. Not the Catalan numbers.

**Problem Set 2**
1. $(i)$ $A = \{3, 5, 7, 9\}$
   $(ii)$ $A = \{3, 5, 7, \ldots, 199\}$ or
   $A = \{\, x \mid x = 2y+1,\ y \in \mathbb{Z},\ 1 \le y \le 99\,\}$
   $(iii)$ $A = \{\, x \mid x = 2y+1,\ y \in \mathbb{Z}\,\}$
   $(iv)$ $A = \{\, x \mid x = 4y,\ y \in \mathbb{Z}\,\}$
2. $(i)$ True  $(ii)$ True  $(iii)$ False
   $(iv)$ True  $(v)$ False
3. $(i)$ $\{a, b, c, d, e, \{a, d\}\}$  $(ii)$ $\{b\}$
   $(iii)$ $\{c, \{a, d\}\}$  $(iv)$ $\{a, b, \{a, d\}\}$
   The sizes are 6, 1, 2, 3.

**4.** Suppose $x \in A$. Since $A = A \cap B$, $x \in A \cap B$ so $x \in B$. Hence $A \subseteq B$.

The converse is: $A \subseteq B$ implies $A \cap B = A$.

**6.** Take $x \in X \setminus (A \cup B)$ and show that $x \in (X \setminus A) \cap (X \setminus B)$ and *vice versa*.

**8.** With three sets you can achieve 8 regions and the maximum number of regions you get using four sets is 14. But you would need 16 regions to depict all possible Venn diagrams involving four sets.

## Problem Set 3

**1.** One-to-one but not onto.

**2.** The first set is a function, but not one-to-one. The second and last sets are one-to-one functions. The third and fourth sets are not functions.

**3.** (*i*) One-to-one.   (*ii*) Not one-to-one.

**4.** (*i*) The function $g \circ f : A \to C$ takes the values $x$, $y$, $x$ and $t$ at 1, 2, 3 and 4, respectively. That is, it can be described by the 4-tuple $(x, y, x, t)$.

(*ii*) The function $h \circ f : A \to A$ corresponds to the 4-tuple $(1, 1, 1, 2)$. The function $f \circ h : B \to B$ is given by $(f \circ h)(a) = c$, $(f \circ h)(b) = b$, $(f \circ h)(c) = b$, $(f \circ h)(d) = a$.

**5.** (*i*) The two one-to-one correspondences between $A$ and $B$ are

$$1 \longleftrightarrow a \qquad 1 \longleftrightarrow b$$
$$2 \longleftrightarrow b \qquad 2 \longleftrightarrow a$$

(*ii*) The six one-to-one correspondences between $A$ and $B$ are

$$1 \longleftrightarrow a \qquad 1 \longleftrightarrow b \qquad 1 \longleftrightarrow c$$
$$2 \longleftrightarrow b \qquad 2 \longleftrightarrow a \qquad 2 \longleftrightarrow b$$
$$3 \longleftrightarrow c \qquad 3 \longleftrightarrow c \qquad 3 \longleftrightarrow a$$

$$1 \longleftrightarrow a \qquad 1 \longleftrightarrow b \qquad 1 \longleftrightarrow c$$
$$2 \longleftrightarrow c \qquad 2 \longleftrightarrow c \qquad 2 \longleftrightarrow a$$
$$3 \longleftrightarrow b \qquad 3 \longleftrightarrow a \qquad 3 \longleftrightarrow b$$

**6.** (*i*) Need to show that $g \circ f$ is one-to-one.

(*ii*) The permutation $g \circ f$ is even if both $f$ and $g$ are odd or both are even; otherwise it is odd.

HINTS AND ANSWERS  139

7. (i) The elements of $\mathcal{A}$ are $(x_0(x_1(x_2x_3)))$, $(x_0((x_1x_2)x_3))$, $((x_0(x_1x_2))x_3)$, $(((x_0x_1)x_2)x_3)$, and $((x_0x_1)(x_2x_3))$.
The set $\mathcal{B}$ is $\{((())), (())(), ()(()), (())(), ()()()\}$.
(ii) $|\mathcal{A}| = |\mathcal{B}| = 5$.
(iii) $f$ is not one-to-one.

8. Hint: from a balanced string of $n$ left and $n$ right brackets construct a sequence by writing numbers in place of the left brackets. Begin at the left with 1 and increase the number you are writing by 1 each time you pass a right bracket. $|\mathcal{F}_n|$ is the $n$-th Catalan number.

**Problem Set 4**

1. (i) $A \times B = \{(1,a), (1,b), (1,c), (2,a), (2,b), (2,c)\}$
(ii) $\{(1,1), (1,2), (1,3), (1,4), (2,2), (2,3), (2,4), (3,3), (3,4), (4,4)\}$.
(iii) $D = \{(1,1), (2,2), (3,3), (4,4)\}$ and $|D| = 4$. There are 24 one-to-one correspondences between $D$ and $A$.

2. 676

3. 6561

4. (i) 17576.    (ii) 46656.    (iii) 33696.

5. $A \times B = \{(a,e), (a,f), (b,e), (b,f), (c,e), (c,f), (d,e), (d,f)\}$.
Since $|A \times B| = |S| = 8$, there are 8! such one-to-one correspondences.

6. $2 \times 26^3 \times 10^3$.

7. (i) 120    (ii) 240

8. Yes.

9. Show that $\sum_{a \in A} |E(a)| = |E|$ and similarly for the other sum.

**Problem Set 5**

1. (i) $3^{10}$
(ii) There are $3^{10}$ functions from a set of 10 students to a set of 3 markers.

2. $2^8$

3. $4_{(3)}$, i.e. 24

4. 136080

5. (i) 2600    (ii) $550!/50!$

6. (i) 499    (ii) 96

7. (i) $11^4$ (ii) 7920

8. 155

9. (i) 8! (ii) $5!\,6_{(3)}$

10. Each permutation $f: C_n \to C_n$ is completely determined by its values at $1, 2, \ldots, n$ and so there are $2^n\, n!$ of them.

**Problem Set 6**

1. Use the binomial theorem to expand the expressions.

2. (i) $\frac{52!}{47!\,5!}$ (ii) $\frac{51!}{47!\,4!}$
   (iii) $\frac{52!}{32!\,5!\,5!\,5!\,5!}$ (iv) $\frac{52!}{32!\,5!\,5!\,5!\,5!\,4!}$

3. (i) 15 (ii) 126

4. (i) $\frac{52!}{39!\,13!}$ (ii) $\frac{52!}{13!\,13!\,13!\,13!\,4!}$

5. $\binom{12}{7}$

6. $\binom{14}{9}$

7. $\binom{m+n-1}{n}$. Consider the solutions to
$$(x_0 - 1) + x_1 + \cdots + x_n = m - 1.$$

8. $2^{n-1}$. Consider a one-to-one correspondence in which the sequence $(k_1, k_2, \ldots, k_r)$ corresponds to the set $\{k_1, k_1 + k_2, \ldots, k_1 + \cdots + k_{r-1}\}$.

**Problem Set 7**

1. (i) 22 (ii) Not possible. (iii) Not possible.

2. 540

3. 20

4. (i) 3 (ii) 6

5. 55

6. $\binom{5}{1}\binom{49}{4} - \binom{5}{2}\binom{48}{3} + \binom{5}{3}\binom{47}{2} - \binom{5}{4}\binom{46}{1} + \binom{5}{5}\binom{45}{0}$

7. (i) 15 (ii) 76

8. $44 = \binom{9}{6} - 4\binom{5}{2}$

9. 31

10. $\sum_{i=0}^{n}(-1)^i \binom{n}{i}(n-i)^m$

HINTS AND ANSWERS   141

**Problem Set 8**  1.  (i)  $\dfrac{9!}{2!\,3!}$   (ii)  $\dfrac{13!}{3!\,2!}$   (iii)  $\dfrac{15!}{2!\,3!\,2!\,2!}$
   (iv)  $\dfrac{17!}{4!\,2!\,2!}$   (v)  $\dfrac{12!}{3!\,2!\,2!\,2!}$   (vi)  $\dfrac{11!}{4!\,4!\,2!}$

2.  $\dfrac{20!}{3!\,4!\,4!\,2!\,3!\,4!}$

3.  3

4.  $\dfrac{15!}{5!\,3!\,4!\,3!}$

5.  $\dfrac{30!}{4!\,2!\,8!\,5!\,4!\,7!}$

6.  (i)  12   (ii)  $\dfrac{7!}{2!\,3!\,2!}$

7.  Put $x_1 = x_2 = \ldots = x_m = 1$ in the Multinomial Theorem.

8.  6

9.  (i)  The multinomial coefficient.   (ii)  $\dfrac{(2n)!}{2^n\,n!}$

**Problem Set 9**  1.  $yz' \vee x(y' \vee z)y$ and $xy' \vee (z \vee x'z')y \vee yx$
2.  $xyz' \vee xy'z \vee xy'z' \vee x'yz' \vee x'y'z'$
3.  $xyz' \vee xy'z \vee x'yz$
4.  Hint: Write $xyz \vee xyz' \vee xy'z \vee x'yz \vee x'y'z' \vee x'y'z$ as

$(xyz \vee xyz') \vee (x'y'z' \vee x'y'z) \vee (xyz \vee xy'z \vee x'yz \vee x'y'z)$.

**Problem Set 10**  1.  A Boolean expression for this function is $x(z' \vee y') \vee x'y$.
2.  A Boolean expression for this function is $y \vee x'z$.
3.  (i)  $x \vee z$   (ii)  $w'y \vee z' \vee x'y' \vee xy$
   (iii)  $z \vee xy' \vee x'y$   (iv)  $yz' \vee w'z \vee wx'y' \vee w'x'y$

**Problem Set 11**  1.  Use truth tables.
2.  (i)  False   (ii)  True   (iii)  True
3.  Use truth tables.
4.  Use truth tables.
5.  Use truth tables.
6.  A tautology.

7. Let the universal set $U$ be the set of all crocodiles $x$. Suppose $H(x)$ mean $x$ is hungry and $M(x)$ mean $x$ is amiable. There are different ways of writing the propositions in symbolic form. Here is one of them:
   - (i) $(\forall x)(H(x) \Rightarrow \sim M(x))$
   - (ii) $(\exists x)(\sim H(x) \Rightarrow M(x))$

8. (i) False. For example, take $x = 0$.
   (ii) True. For example, take $x = 1$.
   (iii) False. For example, take $x = -1$.
   (iv) True. For example, take $x = 1$.

Hint: use the disjunctive normal form.

9. Compound propositions for the 16 truth tables can be given by:

$$\sim p \vee p, \quad p \vee q, \quad p \vee \sim q, \quad \sim p \vee q, \quad \sim(p \wedge q), \quad p, \quad q,$$
$$(\sim p \vee q) \vee (p \wedge q), \quad (p \vee q) \wedge (\sim(p \wedge q)), \quad \sim q, \quad \sim p,$$
$$p \wedge q, \quad p \wedge \sim q, \quad \sim p \wedge q, \quad \sim(p \vee q), \quad \sim p \wedge p.$$

## Problem Set 12

1. Easy!
2. $xy \vee y'z$
3. Use the equivalent expression $x' \vee y$.
4. (i) Suppose $x \downarrow y$ means $x$ **nor** $y$. Then $x' \equiv x \downarrow x$, $x \vee y \equiv (x \downarrow y) \downarrow (x \downarrow y)$ and $x \wedge y = (x \downarrow x) \downarrow (y \downarrow y)$.
   (ii) Suppose $x \mid y$ means $x$ **nand** $y$. Then $x' \equiv x \mid x$, $x \vee y \equiv (x \mid x) \mid (y \mid y)$ and $x \wedge y = (x \mid y) \mid (x \mid y)$.

## Problem Set 13

1. Use the standard methods of mathematical induction.

## Problem Set 14

1. (i) $G(z) = (1+z)^n$  (ii) $G(z) = \dfrac{1}{1-az}$

2. $G(z) = \dfrac{1-4z}{1-5z+6z^2}$

3. $G(z) = \dfrac{2z^2 + z^3}{1 - 2z - z^2 + 2z^3}$

4. (i) $x_n = \frac{1}{2} + \frac{1}{2} \cdot 3^n$  (ii) $x_n = \frac{1}{2}n(n+1)$
   (iii) $x_n = (-1 + 2n)5^n$  (iv) $x_n = \frac{5}{2} - 2 \cdot 2^n + \frac{1}{2} \cdot 3^n$

5. $G(z) = \dfrac{1}{1-z} + \dfrac{4z}{(1-z)^2}$

6. $G(z) = \dfrac{1}{1-z-z^2} + \dfrac{z^3}{(1-z)^2(1-z-z^2)}$

## Problem Set 15

1. Direct substitution!
2. $S_{n+1} - S_n = (n+1)^2$
3. (i) $x_n = 2 \cdot 7^n$ (ii) $x_n = (-1+2n)5^n$
   (iii) $x_n = \frac{5}{2} - 2 \cdot 2^n + \frac{1}{2} \cdot 3^n$
4. (i) $x_n = \frac{1}{2} + \frac{1}{2} \cdot 3^n$ (ii) $x_n = \frac{1}{2}n(n+1)$
5. (i) $x_n = 2 \cdot 2^n - 3^n$ (ii) $x_n = -4 \cdot 1^n + 4 \cdot 2^n - n \cdot 2^n$

## Problem Set 16

1. $\Sigma^2 = \{xy \mid x \in \Sigma, y \in \Sigma\}$, $\Sigma^3 = \{xyz \mid x \in \Sigma, y \in \Sigma, z \in \Sigma\}$, $|\Sigma^2| = 5^2$, $|\Sigma^3| = 5^3$. In $\Sigma^*$, there are $5^5$ strings of length 5.
2. $5^4$
3. $L_1 L_2 = \{abbc, abcd, acbc, accd, debc, decd\}$, $L_1 L_1 = \{abab, abac, abde, acab, acac, acde, deab, deac, dede\}$
4. (i) $\{a, bc, cd\}$ (ii) $\{b, c, cd, cdd, cddd, \ldots\}$
   (iii) $\{ab, cd\}^*$ (iv) $\{dd\}^*$
   (v) $\{ab\}^* \{db\}^*$ (vi) $\{ddd\}^*$
5. (i) $a + b^*$ (ii) $abc(abc)^*$
   (iii) $a^* b$ (iv) $a + b + c + (ab)^*$

## Problem Set 17

1. (a) (i) Not accepted (ii) Not accepted
       (iii) Accepted (iv) Accepted
       (v) Accepted (vi) Not accepted
   (b) (i) No (ii) Yes (iii) No

2. (a) (i) Accepted (ii) Not accepted
       (iii) Accepted (iv) Accepted
       (v) Accepted (vi) Not accepted
   (b) (i) No (ii) No (iii) Yes

3. (ii) (a) Not accepted (b) Accepted
        (c) Accepted (d) Not accepted
   (iii) $L = \{b, ab\}\{b, ab\}^*$ (iv) $(b+ab)(b+ab)^*$

**Problem Set 18**  1. (*i*)  $\{01\}^*$    (*ii*)  $\{aaa\}^*$

2. (*i*) $G = \{N, T, S, P\}$, $N = \{S\}$, $T = \{p\}$ and $P = \{S \to ppS,\ S \to \varepsilon\}$.
   (*ii*) $G = \{N, T, S, P\}$, $N = \{S, D\}$, $T = \{p, q, r\}$, $P = \{S \to Dr,\ D \to pDq,\ D \to pq\}$.

3. $\{abc, a^2b^2c, a^3b^3c, \ldots\}$

4. $\{a^n b^n c^k \mid n \geq 0,\ k \geq 1\}$

**Problem Set 19**  1. (*i*) Use formula (19.1).
   (*ii*) Use formula (19.1).
   (*iii*) Use induction.

# Index

## A

Accepted strings, 115
Accepting states, 114
Addition principle, 25, 29
Algebra,
  Boolean, 58, 63
  of generating functions, 94
  of regular expressions, 86
  of set theory, 11
Alphabets, 108, 115, 124
  input, 115
Alternating sum, 40
An Tu Ming, 5
Argument by contradiction, 82
Arrangements (see Selections), 32, 44
  of indistinguishable objects, 54
  ordered, with repetition, 32
  ordered, without repetition, 33
Arrow diagrams, 17
Asymptotic formulae, 3
Answers, 137
Automata,
  deterministic finite, 114
  non-deterministic finite, 119
Automaton, 114

## B

Balanced strings, 1, 131
Basic connectives, 75
Bijective function, 19, 34
Binomial, 37
  coefficients, 37
  identities, 39
  theorem, 38
Boole, George, 58
Boolean algebra, 58
  laws of, 63
Boolean expressions, 61
  disjunctive normal form, 62
  equivalent, 62
Boolean functions, 59

## C

Carroll, Lewis, 83
Cardinality of a set, 10
Cartesian product, 27
Catalan, Eugène, 5
Catalan numbers, 5, 130
  formula for, 131
  generating function of, 94
  recurrence relation for, 97, 133
  solution to the wagon problem, 21
Characteristic equation, 104, 105
Circuits,
  switching, 60, 64, 70
  digital, 85
  integrated, 85
Closed form, 95
Complement of a set, 10
Composition of functions, 20
Compound propositions, 75
  equivalent, 77
Conclusion, 81
Connectives (logic), 75
Contradiction, 78, 82
Contrapositive, 82
Converse, 82
Counting in two ways, 31, 35
  binomial theorem, 38
  multiplication principal, 29
  number of subsets, 39
  Pascal's triangle, 40
  Vandermonde, 41
  multinomials, 53
Counting principles
  addition, 25
  multiplication, 26

## D

Dead-end states, 115
Decision tables, 59
De Morgan, Augustus, 14
De Morgan's laws, 14, 81
Deterministic finite automata, 114
Difference equation, 101

Digital circuits, 85
Digital logic, 85
Disjoint sets, 25
Disjunctive normal form, 62, 67
Distributive law,
  for sets, 12
  Boolean algebras, 63

## E

Empty,
  function, 33, 39
  set, 8
  string, 5, 108
Equality of sets, 9
Equivalent Boolean expressions, 62, 70
Equivalent compound propositions, 77
Equivalent regular expressions, 111
Exclusive or, 25, 29, 86
Exclusive nor, 86
Expressions,
  Boolean, 62, 67
  equivalent, 111
  regular, 110, 119

## F

Factorial, 34
  falling factorial, 34
Fibonacci numbers, 101, 105, 112, 127
Finite automata,
  deterministic, 114
  non-deterministic, 119
Finite state machines, 114
Formal languages, 108
Functions, 16, 29
  arrow diagram, 17
  bijection, 19, 34
  composition of, 20
  Boolean, 59
  domain, 16
  empty function, 33, 39
  formal definition of, 29
  generating, 94
  image, 16
  injective, 17, 34
  number of, 32
  one-to-one, 17, 34
  one-to-one correspondence, 19
  onto, 18, 50
  surjective, 18, 50
  transition, 115
  well-defined, 22

## G

Gates
  logic, 85
  basic digital logic, 86
Generating functions, 94
  algebra of, 94
  closed form, 95
  for Catalan numbers, 94, 135
  partial fraction expansion, 98
Grammar, 124
  for a language, 126
  language of a, 124
  productions for, 124
Graph, 130

## H

Hints, 137
Hypothesis, 81

## I

Identities,
  binomial, 39
  for sets, 11
  Vandermonde, 41
Image (of a function), 16
Implication, 75, 79
Inaccessible states, 115
Inclusion-Exclusion Principle, 46
  general, 49
Inclusive or, 25, 76,
Induction, 89
Initial conditions, 104
Initial state, 114
Injective function, 17, 34
Inverter, 85
Intersection of sets, 10

## K

Karnaugh map, 67
  method, 68
Kleene, Stephen C, 109
Kleene's theorem, 118

## L

Language, 108
  accepted by a machine, 116, 119
  concatenation, 109
  grammar for, 126
  of a grammar, 124
  non-regular, 111, 120, 125
  regular, 110
  ∗-closure, 109
  union, 109
Linear combinations, 103
Logic, 74
  basic connectives of, 75
  digital, 85
  gates, 85
  of propositions, 74
Logical
  argument, 81
  puzzles, 83

## M

Mapping, 16
Mathematical induction, 89
Multiplication principle, 26, 29
Multinomial coefficients, 54
Multinomial theorem, 55
Multinomials, 55

## N

Negation, 75, 81
Non-deterministic finite automata, 119
  language of, 120
Non-regular language, 111, 118, 125
Non-terminal symbols, 124
Notation,
  $\mathbb{R}, \mathbb{N}, \mathbb{Z}$, 7
  $\in, \notin$, 7
  $\varepsilon$, 5, 108
  $\emptyset$, 8, 109
  $\subseteq, \not\subseteq$, 8
  $\cup, \cap, \setminus$, 10
  $\wedge, \vee$, 61, 75, 86
  $\sim, \Rightarrow, \Leftrightarrow$, 75
  $\exists, \forall$, 78
  ∗, 109
Number of,
  all functions, 32
  bijective functions, 34
  one-to-one functions, 34
  ordered arrangements, 34
  permutations, 35
  regions in the plane, 92, 97

## O

One-to-one correspondence, 19
One-to-one functions, 17, 34
Onto functions, 18, 50
Ordered arrangements, 34
Ordered pairs, 17, 27

## P

Pairwise disjoint sets, 25
Partial fractions, 98
Parity, 20
Pascal, Blaise, 40
Pascal's triangle, 40
Paths (in a DFA), 115
Permutations, 3, 19, 35
  odd and even, 20
  parity, 20
Pigeonhole principle, 18, 121
Predicate calculus, 79
Prefix of a string, 130
Principle,
  addition, 25, 29
  inclusion-exclusion, 46, 49
  multiplication, 26, 28, 29
  of induction, 89
  pigeonhole, 18
Productions, 4, 124
Propositional calculus, 74, 109
Propositions, 74
  compound, 75
  equivalent, 77
  logic of, 74

## Q

Quantifiers, 78

## R

Rational functions, 98
Recurrence relation, 97, 101
  for Catalan numbers, 3, 97, 133
  characteristic equation for, 104, 105
  first order, 102
  general solution, 104, 106
  homogeneous, 101
  initial conditions for, 104
  linear, 101
  second order, 103

Regular expressions, 110, 119
    algebra of, 112
    equivalent, 111
Regular language, 110

## S

Selections, 32
    ordered, 32
    unordered, with repetition, 41
    unordered, without repetition, 37
Sets, 7
    algebra of, 11
    cardinality, 10
    Cartesian product. 27
    complement, 10
    De Morgan's laws, 14
    disjoint, 25
    elements of, 7
    empty, 8
    equality, 9
    intersection, 10
    laws of, 11
    membership, 7
    power, 9
    size, 10
    specification, 7
    subsets, 8
    subsets of size $k$, 37
    union, 10
    universal, 79
    Venn diagrams, 13
Sigma notation, 2
States,
    accepting, 114
    dead-end, 115
    inaccessible, 115
    initial, 114
Stirling's approximation, 3
Strings, 4, 17
    accepted, 115
    balanced, 1, 131
    concatenation of, 109
    empty, 5, 108
    juxtaposed, 109
    prefix, 130
    of a language, 108
Surjective function, 18, 50
Switching circuits, 60, 64, 70
Symbols,
    start, 5, 124
    strings of, 17, 32, 108
    terminal, 124
    non-terminal, 124

## T

Tautology, 77
Terminal symbols, 124
Tree, 130
    planted planar, 130
    vertex, 130
Transition function, 115
Truth tables, 75, 86

## U

Union,
    of languages, 109
    of sets, 10
Universal set, 79

## V

Vandermonde, A, 41
Vandermonde identity, 41
Venn diagrams, 13
Venn, John, 13
Vertex of a tree, 130

## W

Well-defined, 22
Well-formed formulae, 108

## Y

Yang Hui's triangle, 40

## Z

Zhu Shi-Jie, 41